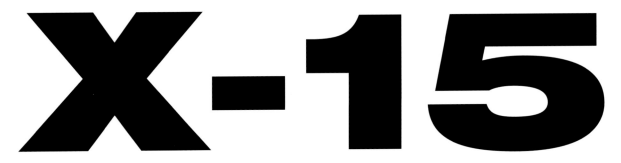

X-15

THE WORLD'S FASTEST ROCKET PLANE AND THE PILOTS WHO USHERED IN THE SPACE AGE

JOHN ANDERSON AND RICHARD PASSMAN

ZENITH PRESS

First published in 2014 by Zenith Press, a member of Quayside Publishing Group,
400 First Avenue North, Suite 400, Minneapolis, MN 55401 USA

All photographs are from the author's collection unless noted otherwise.

Zenith Press titles are also available at discounts in bulk quantity for industrial or sales-promotional use. For details write to Special Sales Manager at MBI Publishing Company, 400 First Avenue North, Suite 400, Minneapolis, MN 55401 USA.

To find out more about our books, join us online at www.zenithpress.com.

Library of Congress Cataloging-in-Publication Data

Anderson, John, 1937-
X-15 : the world's fastest rocket plane and the pilots who ushered in the space age / John Anderson, Richard Passman.
pages cm — (Smithsonian series)
Summary: "The X-15, which flew from 1959-1970, is still the most advaanced research aircraft ever developed and flown, and hangs in a place of honor in the Smithsonian's Air and Space Museum. Its test pilots not only reached the edge of space, but their skill and daring helped engineers understand hypersonic speed and thus pave the way for the Space Shuttle"—Provided by publisher.
ISBN 978-0-7603-4445-3 (hardback)
1. X-15 (Rocket aircraft)—History. 2. Aerodynamics, Hypersonic—Research—United States—History. I. Passman, Richard. II. Title. III. Title: World's fastest rocket plane and the pilots who ushered in the space age.
TL789.8.U6X513 2014
629.133'38--dc23

2013036904

Editor: Elizabeth Demers
Design Manager: James Kegley
Designer: Chris Fayers
Cover Designer: Simon Larkin

On the front cover: *NASM*
On the back cover: *USAF, Air Force Flight Test Center History Office, Edwards Air Force Base*

Printed in China

CONTENTS

EFACE

X-15 is exciting, even on paper. A
cket-powered airplane, the X-15 carries
ly 90 seconds' worth of fuel. It must
aloft under the wing of a B-52 bomber
itude of 45,000 feet, then dropped at a
mber of 0.8. Falling at the acceleration
y (32.2 feet per second, every second), the
t must engage the engine. Then, using
ed fuel supply, he climbs to a maximum
of 352,400 feet or accelerates to Mach
imes the speed of sound). He returns
ds Air Force Base, perhaps 300 miles
decelerating without power through
ic Mach 6 and 5, supersonic Mach 4,
, and then the sonic speed of 1 before
he aircraft to land at 200 miles per hour.
e time of fight from drop to touchdown
esert floor usually lasts about 10 minutes.

The X-15 was a research aircraft, part of a
program designed to study the problems of
hypersonic flight. For that purpose, the plane's
Mach number range had to be above 5, and its
practical altitude range likewise had to be
between 100,000 feet to about 350,000 feet,
above which not enough of the sensible
atmosphere exists to exert a useful aerodynamic

effect. There is no sudden change in airflow
characteristics in progressing faster from
supersonic flow to hypersonic flow as there is
from transonic to supersonic flow, where shock
waves form and a so-called sonic boom occurs
(and where the myth of a "sound barrier" was
created in the 1930s, suggesting that airplanes
could never fly faster than sound). The X-15 was

MACH NUMBERS

Bill Dana and the X-15. *NASA*

Sound moves at a finite speed through air. Its speed depends on the temperature of the air; as the air gets hotter, sound travels faster. At standard sea level conditions, the speed of sound is 340 meters per second, or 761 miles per hour. At 70,000 feet of altitude, where the air is cooler, the speed of sound is 660 miles per hour. Mach number is defined as the ratio of the speed of a vehicle to the speed of sound in the surrounding atmosphere. For example, an airplane flying at a velocity of 4,620 miles per hour at 70,000 feet is flying at seven times the speed of sound, or Mach 7. This leads to the definition of different flight regimes. The subsonic regime is where flight is less than Mach 1. Transonic flight is just below, to just above, Mach 1. Supersonic flight occurs at Mach numbers above 1, and hypersonic flight is considered to be flight at Mach 5 or higher.

designed to explore Mach numbers and altitudes at speeds and heights never previously achieved by manned flight in order to learn about aerodynamic heating, stability, and control. Another purpose was to generate the engineering data that would be gathered for use on future hypersonic aircraft designs, such as the Space Shuttle.

On October 24, 1968, pilot Bill Dana landed the X-15 airplane following what would turn out to be its 199th and last flight. He had reached a Mach number of 5.38 and an altitude of 255,000 feet. The rocket engine had been at 100 percent thrust of 57,000 pounds for 84 seconds, and

the whole flight from launch to touchdown had taken a mere 11 minutes and 28 seconds. This achievement followed nine years of testing that had begun with Scott Crossfield's first unpowered X-15 flight on June 8, 1959, in which the B-52 mother airplane dropped Crossfield and the X-15 at Mach 0.8 and an altitude of 37,550 feet. That flight, strictly a glide flight, lasted only 4 minutes and 57 seconds. From 1959 to 1968, twelve undaunted pilots explored the rocket-powered airplane's performance at hypersonic speeds up to 6.7 times the speed of sound and at altitudes of up to 67 miles (354,200 feet). Their courage and commitment to aviation research and engineering contributed directly to the success of the Space Shuttle program run by the National Aeronautics and Space Administration (NASA).

The X-15—a research aircraft that was never intended to wage war on America's enemies, that was never designed to travel to space, and that was never meant to become an active part of the Air Force fleet—now hangs with distinction in the Milestones of Flight Gallery of the Smithsonian's National Air and Space Museum (NASM). The airplane was donated to the NASM in a ceremony on July 7, 1971, in the Smithsonian Arts and Industries Building. The other remaining aircraft is displayed at the National Museum of the U.S. Air Force at Wright-Patterson Air Force Base in Dayton, Ohio. The third X-15 was destroyed in a fatal wreck, taking with it the life of the pilot, Michael J. Adams.

X-15 at the National Air and Space Museum's Milestones of Flight Gallery. *Photo by Eric Long, NASM*

Several excellent books have been written about the X-15, which the authors wish to acknowledge as valuable reference sources for this book. They are: *At the Edge of Space: The X-15 Flight Program* by Milton O. Thompson (Smithsonian Institution Press, 1992); *Hypersonic: The Story of the North American X-15* by Dennis R. Jenkins and Tony R. Landis (Specialty Press, 2003); *X-15: Extending the Frontiers of Flight* by Dennis R. Jenkins (NASA SP-2007-562, U.S. Government Printing Office, 2007); and *The X-Planes: X-1 to X-45* by Jay Miller (Midland Publishing, 2001). This book is written from a different perspective than these previous titles. It is intended to be a concise biography of the X-15, its mission, and its undaunted pilots and engineers, and will be a companion to the airplane that hangs at the National Air and Space Museum.

We hope that this book can help visitors to the NASM better appreciate why the airplane hangs in the Milestones of Flight Gallery along with such aircraft as the path-breaking Bell X-1 and Lindbergh's *Spirit of St. Louis*.

Finally, the lion's share of the research for, and writing of, this book was carried out by one of the coauthors, Richard Passman, who deserves special acknowledgment for his long hours of work and dedicated effort serving as a volunteer at the NASM, for whom the only reward is the satisfaction and pure joy of living the X-15 experience. Mr. Passman was the chief aerodynamicist for the Bell X-2, the first airplane to fly faster than Mach 3. He shared and contributed to the heady research airplane environment that pervaded the 1950s and '60s, and much of this book reflects his experience in that environment.

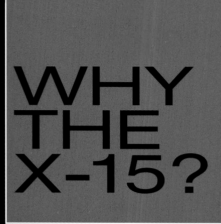

WHY THE X-15?

On October 14, 1947, the Bell X-1 and its Air Force test pilot, Capt. Chuck Yeager, made aviation history by becoming the first to fly faster than the speed of sound. When the sonic boom from this flight reverberated across the desert at Muroc Dry Lake in California, it opened a powerful new chapter in the history of the airplane—the age of supersonic flight. By the early 1950s, supersonic airplanes had become the central focus of airplane design. The Lockheed F-104 Starfighter, for example, was the first airplane designed for sustained cruise at Mach 2. Designed by the famous Lockheed Skunk Works under the direction of iconic designer Kelly Johnson, this airplane exhibits excellent supersonic aerodynamic design. Its fuselage is slender and streamlined, with a pointed nose. Its straight wings are thin, short, and stubby, with a leading edge so sharp that, when on the ground, a protective glove is placed over it to protect the ground personnel from injury—as

Lockheed F-104 Starfighter. *NASA*

F-104 on display at the National Air and Space Museum. *NASM*

well as to protect the leading edge from damage from the ground personnel. These design features were driven by the need to minimize wave drag produced by the shock waves that are present on any supersonic flight vehicle. The thinner the wing and the sharper the nose and leading edges, the weaker are the shock waves, and hence the wave drag is smaller.

The "need for speed" drove the evolution of airplane design. Inspired by the mantra of "flying faster and higher," designers started working in the late 1950s on an airplane capable of flight at Mach 7 and at altitudes higher than 300,000 feet. The result was the first hypersonic airplane, the X-15. One of the three X-15s ultimately produced hangs now in the Milestones of Flight Gallery at the Smithsonian's National Air and Space Museum.

One might think that an airplane designed for Mach 7 would follow the supersonic aerodynamics exhibited by the F-104, but with an even more slender fuselage, a more pointed nose, a thinner wing with a sharper leading edge, etc. However, what we see in the X-15 is a wider fuselage with a blunted nose as well as a thicker wing with blunt leading edges. This is a dramatic departure from good supersonic airplane design because the aerodynamic heating to the aircraft increases with the square of the Mach number, and at

F-104 in flight. *USAF, National Museum of the U.S. Air Force*

SKUNK WORKS

The Skunk Works is a self-contained special projects group within the Lockheed-Martin Company. Its origin within the Lockheed Aircraft Company was in 1944, when Kelly Johnson was given permission to create a small, elite group to design, build, and test the P-80, America's first mass-produced jet fighter. Subsequently, the Skunk Works has become famous for its innovative airplane designs, such as the U-2 spy plane, the SR-71 Blackbird, and the F-117, America's first stealth fighter.

Building the SR-71 at Lockheed's Skunk Works. *Lockheed-Martin*

X-15-2 just after launch, early 1960s. *USAF, Air Force Flight Test Center History Office, Edwards Air Force Base*

hypersonic speeds aerodynamic heating becomes severe. Minimizing the aerodynamic heating becomes the dominant design feature rather than minimizing wave drag. In 1952, H. Julian Allen, then a research engineer at the National Advisory Committee for Aeronautics (NACA) Ames Laboratory in California, proposed the novel idea that aerodynamic heating can be reduced by blunting the nose and leading edges of hypersonic vehicles. The bluntness creates stronger shock waves in front of the vehicle, with higher air temperatures behind the stronger shock waves. The higher air temperatures cause much of the vehicle's potential and kinetic energy to go into the air, leaving less energy to go into the body, thus reducing the aerodynamic heating to the body. (The face of the Apollo lunar return vehicle, which presents itself to the air much like a cannon ball, is a perfect example of the use of a blunt body to minimize aerodynamic heating at hypersonic speeds.)

Indeed, if the super-slender F-104 with its sharp leading edges were to be flown at

Apollo return module. *NASA*

Mach 7, the extreme heat would melt the nose and leading edges, turning them naturally into blunt configurations. This is an example of nature always taking the path of least resistance, blunting the nose and leading edges in a natural attempt to reduce the aerodynamic heating. Designing to minimize aerodynamic heating is the primary reason why the X-15 looks different than the F-104. There are many other aspects of hypersonic flight (arbitrarily defined as flight above Mach 5) that make the X-15 different from a supersonic airplane. The X-15 is a unique airplane in the history of flight and is still today the fastest and highest-flying piloted airplane in existence. The fact that the X-15 is now a museum piece is food for thought.

THE GENESIS OF THE X-15

The first hypersonic vehicles in flight were missiles, not airplanes. On February 24, 1949, a WAC Corporal rocket mounted on top of a captured German V-2 boost vehicle was fired from the White Sands Proving Ground in New Mexico, reaching an altitude of 244 miles and a velocity

of 5,150 miles per hour. After nosing over, the WAC Corporal careened back into the atmosphere at over 5,000 miles per hour, becoming the first object of human origin to achieve hypersonic flight. In this same period, a hypersonic wind tunnel capable of Mach 7, with an 11- by 11-inch cross-section test section, went into operation on November 26, 1947, the brainchild of NACA Langley researcher John Becker. For three years following its first run, this wind tunnel was the only hypersonic wind tunnel in the United States. It later provided key data for the design of the X-15.

The real genesis of the X-15, however, was human thinking, not test facilities. On January 8, 1952, Robert Woods of Bell Aircraft sent a letter to the NACA Committee on Aerodynamics in which he proposed that the committee undertake the study of basic problems in hypersonic and space flight. At that time, several X-airplanes were already probing the mysteries of supersonic flight: the X-1, X-1A, and X-2. Accompanying Woods's letter was a document from his colleague at Bell, Dr. Walter Dornberger, outlining the development of a hypersonic research airplane capable of Mach 6 and reaching an altitude of 75 miles. By June 1952, the NACA Committee on Aerodynamics recommended that the NACA expand its efforts to study the problems of hypersonic manned and unmanned flight, covering the Mach number range from 4 to 10.

After two more years of deliberation, the committee passed a resolution during its October 1954 meeting recommending the construction of a hypersonic research airplane. Among the members of this committee were Walter Williams and Scott Crossfield, who would later play strong roles in the X-15 program. Kelly Johnson, who not only was the Lockheed representative to the committee but was considered to be the country's most famous airplane designer, opposed any extension of the manned research program, arguing that to date

X-15 in flight. *USAF, Air Force Flight Test Center History Office, Edwards Air Force Base*

the research airplane program was "generally unsatisfactory" and had not contributed to the practical design of tactical aircraft. Johnson was the only dissenter; he later appended a minority opinion to the majority report. The spectacular success of the X-15 program and the volumes of hypersonic data it contributed to the design of the Space Shuttle later proved Johnson wrong. The X-15 program was launched.

The X-15 was designed to be, purely and simply, a research vehicle to provide aerodynamic, flight dynamic, and structural response data

Wright Flyer on its first flight at Kitty Hawk (Kill Devil Hills), North Carolina, December 17, 1903. *NASM*

X-15 in captive flight; picture taken from the B-52. *USAF, Air Force Flight Test Center History Office, Edwards Air Force Base*

for use in the development of future manned hypersonic vehicles, such as the Space Shuttle. No hypersonic wind tunnels, past or present, can provide accurate data for the design of a full-scale hypersonic airplane. The frontiers of flight today are the same as they were in the 1950s: the exploration of hypersonic flight. The X-15 will ultimately be viewed as the Wright Flyer of hypersonic airplanes.

The X-15 was the third of a series of research aircraft that were designed specifically to obtain aerodynamic data, beginning with the Bell X-1, the first piloted airplane to fly faster than the speed of sound. The X-1 investigated aircraft behavior primarily in the transonic flight regime. The transonic regime is generally considered to be flight between Mach 0.8 and about 1.3. It begins when air is accelerated to Mach 1 at any local location on the airplane, usually when the airplane is flying at the subsonic airspeed of about Mach 0.8 The second research airplane, the Bell X-1A, investigated supersonic flight to a Mach number of 2.44. This was followed by the Bell X-2, a swept-wing aircraft of stainless steel construction designed to investigate the effects of sweepback and aerodynamic heating to a Mach number of 3.2.

Each of these aircraft, like the later X-15, was rocket-powered and carried aloft to be dropped at an altitude of about 30,000 feet. At these high altitudes, where the air is less dense and the drag is therefore low, the rocket provides maximum acceleration to the airplane following launch. This acceleration is sufficient to allow the airplane to reach the desired speeds and altitudes that allow scientists to study the flight regions between where aerodynamic forces are still useful, and outer space, where they are not, and to study speeds of almost Mach 7, which are solidly in the hypersonic regime.

The X-15 was designed with a very high thrust, 57,000 pounds, provided by an RMI rocket engine with enough fuel for about a minute and a half

INCONEL X

Inconel X is a high-temperature alloy of 72.5 percent nickel, 15 percent chromium, and 1 percent columbium, the rest being iron. It has excellent strength at high temperatures, and it was a natural choice for the X-15 because it could withstand the high surface temperatures expected for the hypersonic flight regime up to Mach 7. Inconel X is a registered trademark of the Huntington Alloy Products Division, International Nickel Company, Huntington, West Virginia.

at full thrust. Researchers wanted to know if the analytical calculations and the wind tunnel data accurately predicted the performance, stability, and control of an airplane flying at Mach 7 at very high altitudes (over 250,000 feet); whether the aerodynamic heating at the high Mach numbers is as high as predicted theoretically; and if the Inconel X structure could maintain its strength at high temperatures.

They also wanted to learn whether the directional stability of the aircraft, which decreases at faster supersonic speeds, could be made sufficient by the X-15's design and by the addition of a stabilization augmentation system (SAS) installed in the airplane. The 199 X-15 test flights evaluated all of these questions.

The risks of flying an airplane designed for testing in an unexplored flight regime are many, both for the known uncertainties and for the unknown. Any research airplane will have a new

single civilian test pilot, Scott Crossfield of the NAA, achieved all the objectives of the program.

Indeed, the X-15 program was as much about the people involved as it was about the data the airplane was designed to collect, or even the airplane itself. The twelve distinguished test pilots who flew this extraordinary aircraft worked hard to learn its characteristics and idiosyncrasies as well as the unknown character of the new flight regime they were investigating. Truly accomplished aviators and apt students of each mission, they bravely addressed each flight with knowledge gained from long hours at flight simulators and with a detailed flight plan. As with any new airplane, difficulties arose. Equipment problems, design unknowns, and other circumstances caused

X-15, rear; XRL-99 rocket. *USAF, Air Force Flight Test Center History Office, Edwards Air Force Base*

X-15 and HL-10 lifting body. *USAF, Air Force Flight Test Center History Office, Edwards Air Force Base*

problems on many flights, although the X-15 flight-testing program claimed only one life in its nine-year history.

The NASA (NACA) flight research crew at Edwards AFB, now known as Dryden Flight Research Center, was a unique and motivated group that built upon their experience with the X-1 airplanes. The first director of the flight research crew for the X-15 was Walt Williams, who was director of the NACA High Speed Research Section, later to become the NASA Flight Research Center. He was also in charge of the early X-1 research flight tests at Pinecastle, Florida. He and his successor, Paul Bickle, ran a rigorous professional organization that continued research begun in the 1920s, when engineers at Langley Memorial Lab wanted to determine the most desirable characteristics for an airplane, as well as innovations in aircraft design that could make flying better, more effective, and safer.

These questions included what data to measure, how to fly to obtain it, how to measure and record it, and, finally, the commitment to publishing this data for the betterment of the industry.

As an example of this research trajectory, the X-15, with Pete Knight at the controls, reached a Mach number of 6.76 on October 3, 1967. On August 22, 1963, the X-15 had gained an altitude of 354,200 feet, more than 67 miles high, with Joe Walker piloting. These incredible achievements were made possible by the use of a supplementary automatic stabilization system, which the successful X-15 test flights proved was necessary in much of the new flight region. Moreover, the X-15 tests also showed that the thermal protection provided by special materials yielded desired favorable results.

The X-15 featured unique design features, including a rolling tail. Each side of the horizontal tail operated separately in opposite directions to roll the aircraft, eliminating the need for ailerons

Dryden Flight Research Center E49-0170 Photographed 10/49
Walter C. Williams, Director of the NACA
High-Speed Flight Research Station. NASA photo

Walter C. Williams. *NASA*

on the wings; ailerons would have induced shock waves at supersonic speeds that would have changed the airflow at the tail surfaces. These shock waves, produced at the deflection hinge lines, would have caused local regions of high aerodynamic heating at that location.

To provide longitudinal control, the two sides of the horizontal tail would operate together in the same direction. The airfoil of the vertical tail surfaces was slab-sided, with a blunt trailing edge; this configuration prevented separated flow on the surface and maintained control at supersonic speeds. The new materials included Inconel X, which maintains its strength at high Mach numbers. Also, the structure was designed

X-15 in flight an instant after drop. *USAF, Air Force Flight Test Center History Office, Edwards Air Force Base*

to minimize the effects of thermal gradients when the outside aircraft skin got hot and the inside stayed cool. The X-15 proved that each of these innovations was successful.

The X-15 was the third and last of a series of air-dropped rocket-powered aircraft designed to investigate high-speed flight regimes from transonic through supersonic to hypersonic velocities. At the time each airplane was conceived and built, there were inadequate wind tunnel or other test data available to assist in the design for flight at these speeds; or in the case of the X-15, the wind tunnel tests had yet to be validated by flight. The X-1, D-558-2, X-2, and X-15 were the first aircraft to fly at Mach 1, 2, 3.2, and 6.7, respectively.

The X-15 had to fly through all the flight regimes that had been pioneered by the earlier research aircraft before extending its speed and altitude range to include the hypersonic regime. These older research aircraft were essentially conventional configurations, with special design

variations required for their specific mission. Like its predecessors, the X-15 had to be dropped successfully from a mother ship, which for the X-15 was the B-52. After drop, the X-15 had to accelerate from subsonic speed through Mach 1 with its attendant shock waves, flow changes, and trim changes. It then climbed and accelerated past the maximum speed of the X-2 to explore the hypersonic regime for which it was designed.

The X-15's rocket engine was a new, much larger version of the RMI rocket engine that powered the X-1 and the X-1A. The new engine needed to increase its thrust from the 6,000 pounds used by the X-1 to the 57,000 pounds required by the X-15's greater Mach number research goals. The new, larger engine was not ready for the early flights, which instead used two of the 6,000-pound engines, combined for 12,000 pounds of thrust. These placeholder engines allowed early flights to proceed, providing data and experience useful for the continuation of the program. The full rocket thrust duration was limited by the quantity of fuel carried and lasted approximately 90 seconds. Since the total time of flight on most missions was about 10 minutes, measured from drop from the B-52 to touchdown on the lakebed at EAFB, this meant that the X-15 flew for about 8½ minutes without any engine power. As in all the rocket research aircraft, the fuel was exhausted in the accelerating portion of the flights so that deceleration, descent, approach, and landing were all performed without power. While the larger X-15 was modified to carry more fuel, this expanded capacity merely extended the plane's speed further into the hypersonic range; it did not provide power for landing.

The X-15 program left an important legacy in the development of manned hypersonic flight. It was, and still is, the fastest, highest-flying piloted airplane in history, and there is no new airplane design being planned in the foreseeable future that could do better. The X-15 met all of its design goals, and the results from its research flights allowed the following, among many others:

1) A verification of existing hypersonic aerodynamic theory and hypersonic wind tunnel techniques
2) A study of aircraft structures under the influence of severe, sustained aerodynamic heating
3) An investigation of stability and control problems associated with acceleration to high altitude, and atmospheric entry at hypersonic speeds
4) A study of the biomedical effects of both weightless and high-acceleration flight

The X-15 was an important steppingstone in the development of the Space Shuttle, which was more space vehicle than airplane but which had to experience hypersonic flight through the atmosphere every time it came back to earth. The spectacular success of the X-15 program is a testimonial to the vision and courage of the engineers and managers who initiated the idea in the first place, the designers who created the vehicle, and the pilots who flew the airplane in the face of many unknowns. It is one of the most important stories in the annals of aviation history in general and aeronautical engineering in particular.

Space Shuttle Columbia, launching. *NASA*

X-15A-2, showing the extra fuel tanks. *USAF, Air Force Flight Test Center History Office, Edwards Air Force Base*

THE ADVENT OF THE RESEARCH AIRPLANE

In 1933, a young aeronautical engineer at the NACA Langley Laboratory conceived the idea of a research airplane that would be designed, built, and flown strictly for the purpose of probing an unknown flight regime. John Stack, a research engineer working in Langley's first high-speed wind tunnel, designed a hypothetical research airplane for the single purpose of collecting data in the subsonic flight regime near the speed of sound. In the early 1930s, little was known and understood about flight near the speed of sound. Because the governing flow equations were mathematically nonlinear in this region, no analytical solutions were available to predict the lift, drag, and stability characteristics for airplanes in this transonic regime. (Even today, the only reliable transonic flow solutions are numerical results obtained from computational fluid dynamics [CFD] using massive supercomputers.) In addition, no accurate transonic wind tunnel data could be obtained from existing high-speed tunnels due to adverse aerodynamic interactions between shock waves from the model, reflecting off the wind tunnel walls and impinging back on the model surfaces.

Bell X-1 in flight. *NASA Dryden Flight Research Center*

Bell X-5 showing swept wings, composite photo. *NASA*

any one or more of the chambers. The X-1 was air-launched from a B-29 bomber; the alternative of taking off from the ground would have consumed too much fuel and not allowed the airplane to reach transonic speeds. Some researchers in the NACA, John Stack included, argued that the research airplane should be powered by a turbojet, thus allowing ground takeoff. Ezra Kotcher and the Army strongly argued against this scenario, and as mentioned earlier, the Army was putting up the money.

Three X-1 aircraft were manufactured by Bell. The first rolled out the Bell factory door on December 27, 1945, without its rocket engine. The unpowered X-1 was transported to the Air Force's Pinecastle Field near Orlando, Florida, for a series of glide tests to examine stability and control characteristics, and to examine low-speed behavior. Carried aloft by a B-29 bomber, the X-1 successfully completed ten glide flights. In each, the airplane behaved beautifully at low speeds. This airplane was then transported back to Bell's factory in Niagara Falls, New York, for installation of its rocket engine. The center of activity now shifted to the Muroc Army Air Field in California, where the powered flights were to take place. There, Bell test pilot Chalmers H. "Slick" Goodlin continued flying the X-1, as called for in the contract. The second X-1 was delivered to Muroc on October 7, 1946, followed shortly thereafter by the first X-1. By May 27, 1947, Bell had completed all the contractually required test flights (all subsonic), and the airplanes were turned over to the Army Air Force.

The Army selected Capt. Charles (Chuck) Yeager to be the next test pilot for the X-1. The Army's first flight, with Yeager at the controls, took place on August 6, when the X-1 was carried aloft by the B-29 carrier aircraft above Muroc for a pilot-familiarization flight. It was the thirty-eighth time that any of the X-1s had taken to the air. Over the next two months the flight-testing program called for a slow increase in speed, gradually approaching the speed of sound. On October 8, Yeager squeezed the airplane to a Mach number of 0.925; two days later, he flew at Mach 0.997. The fiftieth flight took place on October 14, 1947. Although the flight plan did not officially call for it, Yeager brazenly pushed the X-1 through Mach 1, to Mach 1.06. On that day, aviation history was made. It was the first supersonic flight of a piloted airplane, perhaps the most important event in aviation history since the Wright brothers' first successful flight at Kitty Hawk on December 17, 1903. Moreover, the flight was smooth with no technical problems. The existing myth of a "sound barrier" had been broken.

The Bell X-1 lived up to its role as the first airplane designed purely for the acquisition of research data. In total, there were 151 flights, 35 of which were supersonic. The highest Mach number reached by the X-1 was 1.45 on March 26, 1948, with Yeager at the controls. The X-1 was the progenitor of the X-15 in several respects. Both airplanes were rocket-powered. The X-1 proved the viability of a rocket engine for achieving high-speed flight at a time when no other powerplant was available to accomplish the mission. Both were air-launched for the same reason, namely to conserve fuel to enable enough power for a long enough duration to achieve the design Mach number. Ezra Kotcher had argued forcefully for an air launch as opposed to taking off from the ground; he was proven right. This approach carried through to the X-15. The last flight of the X-1 took place on July 31, 1951, piloted by Scott Crossfield, who was also the first pilot to fly the X-15.

Differences in the interests of the three parties involved in the X-1 program were contentious at times. The NACA wanted slow, continuous testing below Mach 1 to fully and safely analyze transonic flow; the Army Air Force wanted to

▲ X-1A in the belly of a B-29 bomber. *USAF*

▼ X-15 and X-1B. *USAF, Air Force Flight Test Center History Office, Edwards Air Force Base*

reach supersonic capability quickly, to develop and build a fighter that would be faster than any enemy; Bell Aircraft wanted to meet its contract requirements and get paid, but also to reach the supersonic flight regime in a timely fashion and thus gain advantage in future procurements. The objectives of all parties were achieved. The NACA did its significant transonic testing and analysis, the Army Air Force had its supersonic airplane, and Bell Aircraft was rewarded for the design, building, and flight testing of the airplane.

THE BELL X-1A

Exactly one month after Chuck Yeager had made history by breaking the sound barrier in the X-1, the Army Air Force began a new study with Bell

for an airplane to fly at Mach 2. Labeled the X-1A, the new airplane had the same wing and horizontal stabilizer and the same rocket engine as the X-1, but it had a completely new fuselage with a more slender shape (higher fineness ratio and increased propellant storage).

On December 12, 1953, Yeager flew the X-1A to a Mach number of 2.44 at 70,000 feet. This set an unofficial world speed record. During the flight, while at this Mach number and altitude, the airplane suddenly encountered inertial roll-coupling and went out of control. Yeager was knocked semi-conscious in the cockpit as the airplane wildly descended. Fortunately, at 25,000 feet, Yeager was able to regain control. Although not intended to be part of the research flight plan, this was the

Douglas D-558-2 Skyrocket. *NASA*

▲ B-52 mother ship taking off with the X-15 mated under its wing. *USAF, Air Force Flight Test Center History Office, Edwards Air Force Base*

▼ X-15 landing with an F-104 chase plane following close by. *USAF, Air Force Flight Test Center History Office, Edwards Air Force Base*

X-15A-2 under the wing of the B-52 mother ship in flight. *USAF, Air Force Flight Test Center History Office, Edwards Air Force Base*

THE X-15 AIRPLANE AND ITS ENGINES

THE AIRPLANE

The X-15 was born on October 5, 1954, when the NACA Committee on Aerodynamics finally decided on the need for a manned hypersonic research airplane. No airplane had even come close to flight at Mach 5 or higher. The Bell X-1 had achieved Mach 1, the Bell X-1A Mach 2.44, and the Bell X-2 Mach 3.2. But to greenlight the development of an airplane that could fly at Mach 7 was truly visionary. No such manned airplane had ever been designed, much less built. Normally, engineers study the previous incarnation of the plane they want to build, innovating from these earlier successful design ideas. But the X-15 was revolutionary—no "before" design even existed. The team would have to start from scratch.

And for good reason.

USAF, Air Force Flight Test Center History Office, Edwards Air Force Base

X-15 with the white ablator coating in flight under the wing of the B-52. *USAF, Air Force Flight Test Center History Office, Edwards Air Force Base*

X-15-3 on the lakebed. *USAF, Air Force Flight Test Center History Office, Edwards Air Force Base*

The X-15 airplane had to be able to accelerate to Mach 7 and climb to over 250,000 feet in order to fill in the unexplored range in speeds above Mach 3.2 and altitudes above 126,200 feet, the maximum achieved by the X-2. (The Bell X-1 had reached 71,902 feet, and the Bell X-1A had reached 90,440 feet.) Like its predecessors, the X-15 would be flown out of Edwards Air Force Base, which was the only installation that had the support equipment and personnel—it was the location of the Air Force Test Pilot School—to

handle the research test flights. Moreover, because of the high landing speed of the X-15, Edwards had the only "runway" long enough for landing the airplane—essentially the whole expanse of the Muroc Dry Lake bed.

The new airplane, like the X-1 and X-2 before it, would be rocket-powered with high thrust, and it would be carried aloft in a "mother ship" to save fuel by applying the thrust at an altitude where the air density was low (hence, low drag). The X-15 would also have to carry enough fuel

▲ X-15 under the wing of the B-52 in flight. *USAF, Air Force Flight Test Center History Office, Edwards Air Force Base*

▼ X-15-1 mounted under the wing of the B-52 before its first flight, June 8, 1959, with Scott Crossfield in the cockpit. *USAF, Air Force Flight Test Center History Office, Edwards Air Force Base*

MONOCOQUE

Monocoque is a French word meaning "single shell." Here, the fuselage is a single, hollow shell that carries on its surface the aerodynamic loads exerted on the fuselage. A monocoque fuselage allows maximum space inside the fuselage for internal components, such as fuel and oxidizer tanks, and electronic equipment. A semi-monocoque structure has additional elements inside the shell, such as formers that conform to the cross-sectional shape of the fuselage and stringers that run longitudinally along the fuselage. These provide additional structural strength while still preserving volume inside the fuselage for other components. These structural elements can be seen in the cutaway view of the X-15A-2 shown on the opposite page.

then through supersonic and hypersonic flight in getting to and from the targeted data points, and finally it would have to decelerate from hypersonic flight to return to the landing site, followed by descent and landing that had to occur at relatively low subsonic conditions.

Because of these specific design requirements, the engineers started with a blank slate, using all of the latest technologies that might apply to the new airplane and the extreme conditions, known and unknown, that it would endure. They also built upon their previous experience and knowledge of the known flight regimes to design an aircraft that could unveil the mysteries of hypersonic flight.

The X-15's fuselage, wings, tail, size, and weight generally look conventional. The fuselage structure is monocoque and semimonocoque. The pilot compartment was a little more ample than that of a fighter jet. The wing has a span of 22 feet, uses an NACA 66005 symmetric laminar flow airfoil, has an area of 200 square feet and an aspect ratio of 2.5, and features a sweepback angle at the quarter chord of 25 degrees. The horizontal tail is tilted down from the fuselage, and the upper vertical tail looks like most others except that the airfoil is wedge-shaped with a blunt trailing edge, unlike the usual airfoil shapes.

But there are two major changes that further distinguish the X-15:

First, there are no ailerons on the wing; roll-control is achieved by deflecting differentially the right and left sections of the horizontal tail. Also, the horizontal tail has no elevators; instead, the whole right and left sections deflect in the same direction together to provide pitch control.

Second, the vertical tail has an unusual airfoil section. It is essentially a vertical slab, small and rounded at the leading edge and flat-sided at a 5-degree half-angle out to the trailing edge, which is blunt.

to allow the high thrust to operate long enough to accelerate to the speeds and altitudes needed to perform the mission. So, the airplane had to be big enough for the fuel volume needed and be able to carry a rocket engine with far more thrust than employed previously, as well as structural materials that would maintain strength at the high temperatures to which the airplane would be subjected at its high speeds of flight.

The design also had to consider the requirements of the nonhypersonic flight regimes for the other portions of flight: It would drop from the mother ship at high subsonic speed, accelerate through Mach 1 and the transonic speed region,

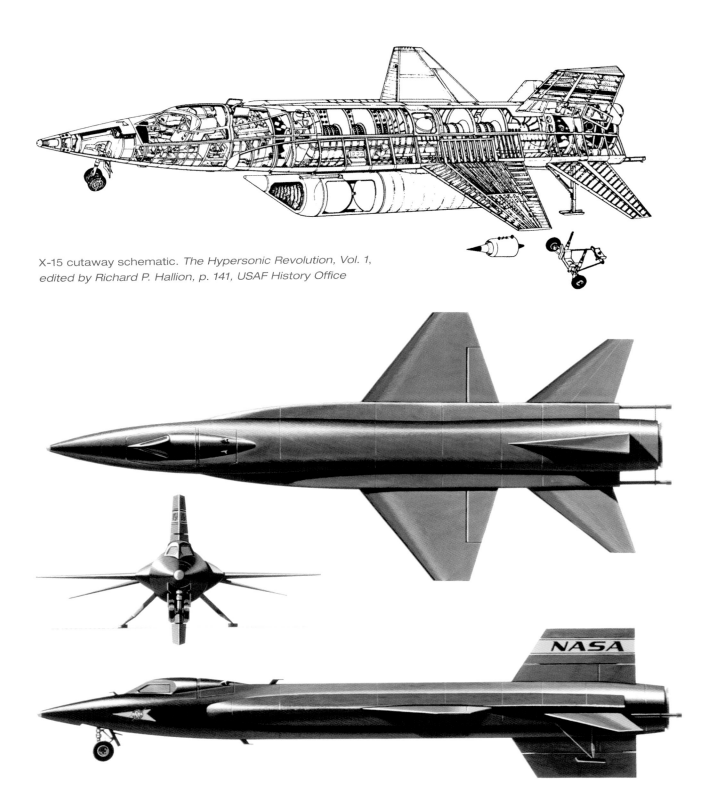

X-15 cutaway schematic. *The Hypersonic Revolution, Vol. 1,*
edited by Richard P. Hallion, p. 141, USAF History Office

X-15 three-side view.
NASA Dryden Flight Research Center

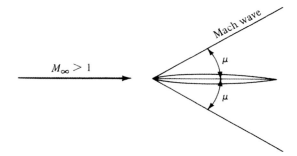

Mach waves (very weak shock waves) on a needlelike body.

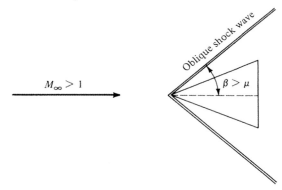

Oblique shock waves on a wedge-type body, demonstrating that the stronger shock wave is at a larger angle than the weak Mach wave.

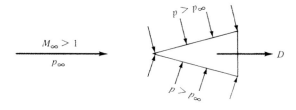

Demonstration of the constant pressure exerted on the face of the wedge, downstream of the shock wave.

The leading edge is rounded in order to reduce the aerodynamic heating in that region. Overall, the vertical tail is a geometrically simple 10-degree total angle wedge with a blunt, flat surface for the trailing edge. The wedge shape has two aerodynamic advantages at supersonic

and hypersonic speeds. First, the pressure on the flat sides is a constant downstream on the nose, and this encourages attached flow over the whole surface all the way to the blunt trailing edge. Expansion waves occur at each corner of the trailing edge. These expansion waves are the direct opposite of shock waves. The pressure decreases through an expansion wave, whereas it increases through a shock wave. The flow leaves the trailing edge through an expansion wave, and hence the pressure on the flat base of the vertical tail is lower. This in turn increases the aerodynamic drag on the vertical tail, called base drag, but at hypersonic speeds the base drag is a very small fraction of the overall drag.

The second aerodynamic advantage, and the primary reason for the use of the wedge shape, is increased directional stability. In August 1954, Charles H. McLellan, head of the 11-inch hypersonic wind tunnel at the NACA Langley Aeronautical Research Laboratory, published some stunning and almost counterintuitive results in NACA Research Memorandum LF44F21 entitled "A Method for Increasing the Effectiveness of Stabilizing Surfaces at High Supersonic Mach Numbers." His work showed that the wedge shape "should prove many times more effective than the conventional thin shapes optimum for lower speed."

The wedge shape took advantage of the nonlinear physics of shock waves as follows: If a surface in a supersonic flow is already inclined at an angle to the flow, say 5 degrees like the surface of a 5-degree half-angle wedge, and then the wedge itself is further inclined by an additional 2 degrees due to a control input, the pressure and hence the aerodynamic force on that surface (which is now at 7 degrees to the flow) is much higher than what would occur on a thin airfoil shape simply deflected by 2 degrees. Aerodynamicists at North American were aware of McLellan's work, and they put this NACA

X-15-3 with ablative coating mounted under the wing of the B-52 in flight. *USAF, Air Force Flight Test Center History Office, Edwards Air Force Base*

X-15A-2, showing the external fuel tanks on the ramp of the NASA Flight Research Center at Edwards. *USAF, Air Force Flight Test Center History Office, Edwards Air Force Base*

unstable without the system. The airplane was designed with reasonable cockpit visibility; the pilot could see all around, but he could not see the wings or the nose of the airplane.

The fuel for the later X-15 flights using the XLR99 engine was anhydrous ammonia, and the oxidizer was liquid oxygen (LOX); the fuel for the earlier flights using the XLR11 engine was water-alcohol. Both fuel and oxidizer were carried in the fuselage and held by the outside structure of the fuselage. The fuselage also contained the hydrogen peroxide (H_2O_2), used for the small control rockets that operated at high altitudes. Nitrogen pressurized the cabin, and helium pressurized the fuel and oxidizer.

A detail showing the X-15 being mounted under the wing of the B-52. *USAF, Air Force Flight Test Center History Office, Edwards Air Force Base*

X-15 in flight. *USAF, Air Force Flight Test Center History Office, Edwards Air Force Base*

THE
TEST
ARENA

For the X-15 program to be a success, the airplane and the pilots had to have a home—a physical facility for servicing the aircraft and a takeoff and landing area. Each flight required teams of support people on the ground as well as other pilots and airplanes in the air. All of these constituted the test arena.

EDWARDS AIR FORCE BASE

The X-15 flight tests occurred at Edwards Air Force Base, located about 100 miles northeast of Los Angeles. It is located on Rogers Dry Lake, a 44-mile-long pluvial lake in the Mojave Desert, which is the world's largest pluvial lake (sometimes called paleolakes because they are caused by heavy rain during periods of glaciation). This dry lake maintains a smooth surface because winds consistently sweep the winter rains back and forth across the lakebed. Most of the year, the lakebed is dry and flat with a variation of height of only about 18 inches from one end to the other.

B-52 in flight with the ablative-coated X-15-3 attached. *USAF, Air Force Flight Test Center History Office, Edwards Air Force Base*

▲ X-15 in flight. *USAF, Air Force Flight Test Center History Office, Edwards Air Force Basea*

▶ X-15 run-up area at Edwards Air Force Base, 1958. *USAF, Air Force Flight Test Center History Office, Edwards Air Force Base*

There are a number of dry lakes in this high desert region, some of which made suitable alternate sites for the emergency landings that might occur, and occasionally did occur, during the flight-testing program. The lakebed had to be smooth enough and hard enough to support an airplane that landed on skids, without digging in and causing an accident, but also long enough for a normal landing. The maximum travel distance from launch to landing was set by the high-altitude flight, where the glide from altitude to landing required a 300-mile distance from launch to Edwards Air Force Base. The alternate fields selected were located within glide range at launch along the path from the launch site to Rogers Dry Lake at Edwards.

The U.S. Army Air Force had used Rogers Dry Lake, then known as Muroc, since the 1930s. During World War II, the Army used the site for flight testing. The advantages of the site include the long, effective runway offered by the lakebed and the 15,000-foot concrete runway that had been built during the war. Other advantages that Rogers afforded were the good weather that enabled many flying days and the security of being essentially in the middle of nowhere, both of which ensured control over the flights. It also provided security for classified aircraft.

While Air Force personnel maintained tight security during the X-1 and X-2 flights, they were more relaxed with the X-15, primarily because it was a research airplane, not intended for combat. Edwards Air Force Base was where all the new military airplanes were tested, including airplanes of super-secret nature, earmarked for eventual combat. Thus, security was at a maximum. By the time of the X-15, however, research airplanes were viewed as just that, research tools. They were thus lower in the hierarchy of security. Most details of the X-15 airplane, the flight tests, and the data were not kept secret. Security for the X-15 was more in the nature of "watchman" and "housekeeping." Those responsible made certain that no unauthorized people had access to the airplane, that tools were not left in the cockpit by accident, etc.

The first U.S. jet airplane, the Bell P-59, was tested on October 2, 1942, at Muroc by Bell's chief test pilot, Bob Stanley. When the X-1 outgrew the initial test site at Pinecastle, Florida, the Air Force selected Rogers Dry Lake for its subsequent flights. There, on October 14, 1947, Chuck Yeager flew the X-1 to the first supersonic flight, reaching a Mach number of 1.06 at 43,000 feet altitude. The NACA High Speed Flight Section under Walter Williams, who was responsible for the X-1 testing, continued in the testing of the Douglas D-558-2 and the Bell X-2 rocket-propelled aircraft, as well as other aircraft flown for test purposes before the creation of the X-15. The site also boasted the presence of the USAF Test Pilot School, whose pilots and aircraft supported the X-15 test flights in many ways, including flying chase aircraft deployed along the X-15 flight path.

The area was known as the high desert because Edwards Air Force Base was at 2,500 feet altitude and the alternate fields ranged up to 5,700 feet. Landing at an altitude higher than sea level requires

DC-3 and C-130 support aircraft at Mud Lake. *USAF, Air Force Flight Test Center History Office, Edwards Air Force Base*

Flyover by the B-52. On the ground are the X-15, Piasecki X-21 helicopter, and ground support personnel and equipment. *USAF, Air Force Flight Test Center History Office, Edwards Air Force Base*

B-52 with the X-15 attached, taxiing before takeoff for its flight on November 3, 1965, with pilot Bob Rushworth in the X-15. *USAF, Air Force Flight Test Center History Office, Edwards Air Force Base*

a longer ground distance, since the air is less dense; thus, speed at landing has to be higher. Decelerating to stopping from a higher speed at landing by necessity requires a longer landing distance.

On November 9, 1962, X-15 pilot John McKay embarked on a routine flight to reach a Mach number of 5.5 and an altitude of 125,000 feet. Though McKay's flight plan called for full power, the engine was putting out only 35 percent power, and ground control directed McKay to shut off the engine and land at Mud Lake, one of the emergency landing sites. McKay jettisoned some of the remaining fuel as required by protocol, but the routine emergency landing was complicated when the flaps didn't deflect downward to increase lift, resulting in a dangerously high-speed landing at 257 knots. This caused a failure to the main landing skid, which in turn caused the left wing and stabilizer to dig into the lakebed, flipping the X-15 upside down.

McKay jettisoned his canopy during this flip-over, but his helmet was the first thing to hit the ground. The rescue crew and the fire truck sped to the airplane. Fumes from the crash prevented them from approaching, but the H-21 helicopter pilot used his rotor blades to blow the fumes coming from the anhydrous ammonia fuel that leaked from the aircraft, so that rescue could proceed. The rescue crew was able to dig the ground out from under McKay and extract him. A C-130 arrived with paramedics and more rescue personnel, and they flew McKay to Edwards Air Force Base before tending to the damaged X-15. The emergency preparation and actions saved McKay's life and showed the crucial importance of alternate fields and the support teams who staffed them.

The X-15 pilots did not want to land at these alternate fields. They were for emergencies only. Landings there were the same as those as at Edwards—dead-stick landings with no power to make adjustments for height or location during landing, nor to abort the landing approach and go around to try again. In his book *At the Edge of Space: The X-15 Flight Program*, Milt Thompson summed up the pilots' preferences:

X-15 after engine failure forced pilot Jack McKay to crash-land upside down at Mud Lake, November 9, 1962.
USAF, Air Force Flight Test Center History Office, Edwards Air Force Base

Rogers (dry lake) was where God intended man to land rocket airplanes. It was big. It had many different runways. It was hard. It had no obstructions in any of the many approach paths. It had all of the essential emergency equipment. It was territory that we were intimately familiar with, and it had a lot of friendly people waiting there. In other words, it was home.

THE B-52 CARRIER AIRCRAFT

The X-15 flights would not have been possible without the B-52A, which carried the airplane under its right wing. Edwards Air Force Base is huge, and it includes the whole of Muroc Dry Lake. Not only did the flights originate at Edwards, both the X-15 and its mother ship, the B-52, landed

there also, although on different plots of ground at the site. The B-52 started on the runway at zero velocity, accelerated to takeoff, and carried the X-15 to its launch position with a speed of approximately M=0.85 and an altitude of about 45,000 feet. While the X-15 achieved a record speed of M=6.7, the first 0.85 was accomplished by the B-52 in the first phase of the flight. The B-52 also sometimes positioned the drop location as far away from Edwards as 300 miles, whereas the flight profile dictated for the X-15 to land at Edwards. The X-15 expended no fuel for such a running start, which was required to obtain the data sought by the test. It took about an hour and a half from takeoff to get to the launch position; the rest of the X-15's flight to its landing was an additional 10 minutes.

Both the X-1 and the X-2 rocket-powered research aircraft were also carried aloft from

X-15 landing with an F-104 chase plane alongside.
USAF, Air Force Flight Test Center History Office, Edwards Air Force Base

ROGERS LANDINGS VS. LANDINGS AT EMERGENCY SITES

Listed below are the number of landings that took place at alternate fields, to be compared with the 188 normal landings at Rogers Dry Lake.

  2 Cuddeback
  1 Delamar
  4 Mud
  1 Rosamond
  1 Silver
  1 Smith Ranch

Since these were emergency fields, they had to have equipment there and personnel on site to act in case they were needed. Prior to the flights, equipment such as a fire truck with 500 gallons of water, a helicopter, firemen, an Air Force pilot to act as the lake controller, an AF crew chief, an AF doctor, an AF pressure-suit technician, and a NASA X-15 specialist were deployed. A test flight was a big operation, and a cancelation was a waste of time for many.

Edwards Air Force Base by carrier or "mother" aircraft, the B-29 for the X-1 and the B-50 for the X-2. The mechanical alterations required to the carrier aircraft were principally in the bomb bay area in order to securely hold the research aircraft and to provide a reliable launch mechanism. The research aircraft pilots rode to the launch altitude and speed in the carrier aircraft, did the checkout before launch within the carrier aircraft, and replaced the liquid oxygen that had boiled off during the climb, all before entering the research airplane. For the X-15, the mother ship was supposed to have been the B-36, and the X-15 would have been carried to its launch position in the bomb bay opening. Some of the reasons the B-52 made the cut instead were related to differences in the availability and cost of each aircraft and the parts required for its maintenance during the flight-test program.

The B-36, then in the process of being phased out as an active bomber in the Air Force inventory, was a maintenance nightmare, whereas the then-modern B-52 was (and still is today) the main bomber for the Strategic Air Command. Moreover, the weight of the X-15 increased during the design phase, and the extra capability of the B-52 could

more easily achieve the speeds and altitudes required by the data regions. Changing from the B-36 to the B-52 meant that the X-15 pilot could not ride inside the carrier aircraft. Using the B-52 meant that the X-15 had to be mounted on a pylon under the B-52's right wing.

There was no way for the pilot to transfer from the B-52 to the X-15 after takeoff, which meant that he had to remain inside the X-15 during takeoff and for the roughly hour-and-a-half climb to position. This increased the pilot's risk significantly. In an emergency during the launch-to-climb phase, the B-52 would have to drop the X-15 and its pilot rather than risk the lives of the entire operation's crew. If the X-15 could be dropped, its pilot could possibly glide to a dry lakebed, or eject if the altitude was high enough. There were a number of captive flights—i.e., while the X-15 was still attached to its mother ship—where problems arose of such a nature that the launch was aborted, such as the auxiliary power unit (APU) not functioning in checkout or electrical signals not transmitting properly. In these circumstances, the B-52 landed safely with the X-15 still tucked under its wing. On such occasions, it must have seemed like a long, fruitless mission for the captive X-15 pilot. Luckily, neither the B-52 nor the X-15 pilots ever had to face such an unplanned drop.

The B-52 required numerous modifications to allow both airplanes to replenish the liquid oxygen, to accommodate the mating of the two aircraft, to assure that the B-52 had adequate control for the mission, and to assure that structural sufficiency was proper for both aircraft. (The X-15's fuel was anhydrous ammonia, which does not boil off and does not require topping off, meaning that only the liquid oxygen required replenishment.) Twenty-seven B-52 pilots supported the X-15 flights. Two of the first were Capt. Charles Bock and Capt. John Allavie.

Top: X-15 mating area. *USAF, Air Force Flight Test Center History Office, Edwards Air Force Base*

Above: X-15 in the process of being mated to the B-52. *USAF, Air Force Flight Test Center History Office, Edwards Air Force Base*

The activities of the B-52 airplanes and their USAF pilots over nine years were integral to the success of the X-15 program. It was not a minor expense.

CHASE AIRCRAFT

Chase aircraft are high-speed aircraft whose pilots observe the physical status of the X-15 during its mission, principally during its climb with the B-52 and then toward the end of the X-15's test flight. They are positioned near alternate landing fields, at approach to landing through touchdown, and during the landing run-out.

During the climb, while the X-15 is attached to the B-52 mother ship, the chase pilot observes the X-15's external features, makes control-surface checks, and observes any irregularities during the climb. In making control-surface checks, the chase pilot observes the physical deflection of the control surfaces, which for the X-15 are the rudder and the horizontal tail, as deflected by the pilot in the cockpit and observed by the pilot in the chase plane. The pilot in the cockpit cannot see these control surfaces, and so it falls to the pilot of the chase plane to observe them. This check is done before the X-15 is dropped from the B-52. It is an essential safety check; if the control surfaces are not working, the flight is scrubbed.

At drop, the chase pilot watches the engine start up, observes the power levels, notes the clearance from the B-52 as the X-15 separates, and

B-52 in flight with the X-15 attached and the F-100 chase plane alongside. *USAF, Air Force Flight Test Center History Office, Edwards Air Force Base*

B-52 in flight with the X-15 attached and the T-38 chase plane alongside. *USAF, Air Force Flight Test Center History Office, Edwards Air Force Base*

is there to assist in descent to landing at Edwards AFB if, for example, the engine doesn't start and the X-15 heads for an emergency landing at the dry lake designated for that particular launch. He can note all the external features of the X-15, its sink rate, its progressive proximity to the ground, and anything unusual that would help the pilot during the landing, such as anomalies in configuration if the flaps did not deploy. The chase pilot can quickly land during an accident in order to physically assist or help rescue the pilot. In emergencies, he would perform the same functions when stationed near alternate landing sites.

With flights varying from launch close to Edwards Air Force Base to launch 300 miles distant, different numbers of chase planes were needed. Usually there were four, one for the climb of the X-15 and the B-52 mother ship, another

at drop, one at an intermediate station above an alternate landing field, and one to cover the descent and landing at Edwards. During the most distant launch, an additional chase plane was needed to cover additional emergency field locations. As a result, there were either four or five chase planes used per X-15 flight. These chase pilots were usually other X-15 pilots, NASA research pilots, or Air Force pilots from the Air Force Flight Test Center.

The chase airplanes that were chosen best matched the X-15's flight characteristics required by the X-15 testing program. For the early flights launched at Edwards Air Force Base, an F-100 answered the call. Later, the team chose a Northrop T-38A because it better matched the B-52's speed during its right turns. Both the F-100 and the T-38A could fly in the low supersonic

▲ Another view of the X-15 landing with the F-104 chase plane alongside. *USAF, Air Force Flight Test Center History Office, Edwards Air Force Base*

▼ Another view of the F-104 chase plane. *USAF, Air Force Flight Test Center History Office, Edwards Air Force Base*

F-104 chase plane. *USAF, Air Force Flight Test Center History Office, Edwards Air Force Base*

range, around Mach 1.5. If there was a problem in climb and cruise to launch, the chase pilot was thus in position to help in the landing.

For the launch at a distance from Edwards Air Force Base, an F-104 chase aircraft stayed with the X-15 until it accelerated out of sight. The F-104 was the first fighter airplane capable of sustained flight at Mach 2. The pilot of the F-104 observed the separation from the B-52 at drop and watched the engine for proper light-up. If the engine did not fire properly, the F-104 would descend with the X-15 to landing and be on hand to help on the ground.

For the chase aircraft covering the intermediate emergency fields, F-104s assisted in the descent and landing of the X-15 and provided any assistance needed after touchdown. These aircraft

B-52 in flight with the X-15 attached and the T-38 chase plane nearby. *USAF, Air Force Flight Test Center History Office, Edwards Air Force Base*

delayed their takeoff for about 30 minutes after the B-52 took off so they would have enough fuel to loiter at their positions.

Other flight vehicles participated as well. A helicopter, the Piasecki H-21, ferried personnel to and from emergency fields as required. It also blew fumes away from damaged aircraft, as when Jack McKay flipped over during his emergency landing. This allowed emergency personnel to extricate him from his airplane and perform other functions during his rescue.

Air Force C-130s transported equipment and personnel to emergency fields, including fire engines. Safety was taken seriously.

X-15 on the lakebed after the flight on October 17, 1961, with pilot Joe Walker still in the cockpit and the Pasecki H-21 helicopter in the background. *USAF, Air Force Flight Test Center History Office, Edwards Air Force Base*

Rear view of the B-52 on the ground with the X-15 attached to its right wing. *USAF, Air Force Flight Test Center History Office, Edwards Air Force Base*

Support trucks and personnel at an X-15 landing site. *USAF, Air Force Flight Test Center History Office, Edwards Air Force Base*

05

THE PILOTS

The X-15 program was a success, thanks in no small part to the men who flew the airplane. Each of the X-15 test flights was an example of intense man-machine interaction, and each of the twelve pilots who flew the X-15 were as finely tuned and technologically sophisticated as the machine itself. They set speed and altitude records for a manned airplane that still stand today, and they pioneered new piloting techniques for hypersonic aircraft that were not only adapted for the Space Shuttle but will continue to be used for future manned hypersonic aircraft. The X-15 pilots were brave and professional, venturing into a totally unknown regime of flight, and they helped to write the book on manned hypersonic flight for the next generation.

Bob White standing beside the X-15. *USAF, Air Force Flight Test Center History Office, Edwards Air Force Base*

All of the X-15 pilots at one time or another were members of the elite NASA Flight Research Center at Edwards Air Force Base. The flight research team was under the direction of Walter C. Williams, who managed a group that planned all the flights, determined what data to acquire, gave the pilots what they needed to obtain the data in an effective and safe manner, and determined how to react in emergencies. Williams and his team were in charge of the flight testing of all the X-airplanes through transonic and supersonic regimes leading up to the X-15, namely the X-1, X1A, D558-2, and the X-2. This center had started out as a small group of about 27 people in 1946 dealing with the X-1 and grew to about 500 at the time of the X-15. These people collectively:

1) Maintained the aircraft, housed, repaired, modified, and prepared the airplane for each flight.

2) Provided for each flight. This included ground crew efforts to ready the airplane, provide the instrumentation, assure the safety for the airplane, provide the chase aircraft and their pilots, and provide emergency gear like the fire trucks and helicopters, as well as the communication links.

3) Provided plans and procedures for each flight, including a detailed pilot checklist for the X-15 and the B-52 mother ship.

4) Provided a flight plan for the X-15 to obtain the requisite data. This sequence included the drop from the B-52, rocket firing and powered flight, climb and transition to level flight, unpowered flight to the speed and altitude required for the data, and finally return to base and landing.

5) Provided a simulation plan to train the pilot for obtaining the data in flight, alternate flight paths to the desired data points if the airplane was over or under the speeds and altitudes planned, and emergency response to various potential problems during the flight. The Flight

Research Center had a special flight simulator designed for the hypersonic regime.

6) Conducted the flights with all the equipment, chase pilots and planes, and communication lines to assist the X-15 pilot to assure safety and performance.

7) Reduced and evaluated the flight data, and utilized the results in future activities.

In September 1959, Walter Williams left the Flight Research Center for the first of many executive positions in the space program, beginning with director of operations for Project Mercury. He was replaced at the Flight Research Center by Paul F. Bikle, who continued Williams's rigorous professional standards. All the important accomplishments of the X-15 program were performed under Bikle.

The first flight of the X-15 took place on June 8, 1959. Carried aloft under the wing of a B-52, the experimental vehicle was released with its pilot at an altitude of 37,550 feet. Unlike all subsequent X-15 flights, however, there was no roar of the rocket engine. Indeed, there were no propellants aboard; this was intended to be a gliding flight, pure and simple. Its purpose was as a familiarization flight, the first checkout of the flight characteristics of the airplane in its glide down to landing, the response to the control system, the stability of the airplane, the handling of the control forces by the pilot, the response rate of the airplane to the controls, and its motion at touchdown and landing.

Nevertheless, the X-15 reached a speed of Mach 0.79 on its maiden descent to the desert floor. Moreover, as with all the other 198 X-15 test flights, a problem occurred. The airplane began to pitch up and down, a longitudinal oscillation that rapidly increased in amplitude. The pitch damper designed to avoid this oscillation was discovered to be inoperable. Fortunately, the X-15 touched down safely at the bottom of an

Six of the twelve X-15 test pilots. From left to right: Joe Engle, Bob Rushworth, Jack McKay, Pete Knight, Milt Thompson, and Forest Peterson. *USAF, Air Force Flight Test Center History Office, Edwards Air Force Base*

oscillation, suffering damage only to the landing gear. A. Scott Crossfield, the pilot who had the most influence of all the X-15 pilots on the design and flight performance of the airplane, performed the difficult maneuver. In all other aspects, the plane performed as anticipated by the designers.

SCOTT CROSSFIELD

1921–2006

Scott Crossfield was more than just the first man to fly the X-15; he was the only one of the twelve test pilots who contributed directly to the airplane's design and to the design of its flight-test program. Crossfield successfully combined his master's degree in aeronautical engineering with his exceptional piloting ability and experience to enhance the design and operation of an experimental vehicle that would go far beyond the known atmospheric flight spectrum, to speeds of almost Mach 7 and to altitudes higher than 350,000 feet.

Scott Crossfield was born on October 2, 1921, in Berkeley, California, and attended college at the University of Washington in Seattle, beginning in 1940. The outbreak of World War II interrupted

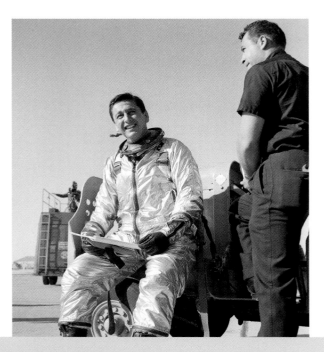

Scott Crossfield in his pressure suit for a preflight briefing. *USAF, Air Force Flight Test Center History Office, Edwards Air Force Base*

Crossfield in the X-15 cockpit. *USAF, Air Force Flight Test Center History Office, Edwards Air Force Base*

his studies in 1942, when he joined the Navy. After he received his pilot's wings and ensign's commission in 1943, the Navy assigned him to be a flight instructor and maintenance officer. He served in the South Pacific for six months but did not see combat duty. His piloting skills put him at the helm of a Navy aerobatic team, and he flew Corsair fighters for a short period following the war. Crossfield was, however, an aeronautical engineer at heart, and he returned to the University of Washington in 1946 to finish his bachelor's degree in aeronautical engineering, as well as his M.S., in 1949. During that time, he obtained valuable experience working in the Kirsten Wind Tunnel at Washington.

It was not a good time to graduate with an aeronautical engineering degree; the industry was suffering from large government cutbacks in defense after World War II. However, the advent of the Korean War in 1950 reversed this situation, and suddenly the aircraft industry was back on its feet. Crossfield found a position as an aeronautical research pilot with the NACA High Speed Flight Station (now the NASA Dryden Flight Research Center) at Edwards Air Force Base in June 1950. The time and opportunity were ripe for Crossfield; over the next five years, he was to fly virtually all the experimental airplanes at Edwards, including the Bell X-1, the delta-wing XF-92, the X-4, the X-5, and the Douglas D-558-1 Skystreak. On November 20, 1953, he became the first person to fly at Mach 2 while piloting the rocket-powered Douglas D-558-2 Skyrocket to a speed of 1,291 miles per hour in a shallow dive.

DOUGLAS D-558-2

Powered by a rocket engine, and developed by Douglas for the U.S. Navy, the Douglas D-558-2 explored transonic and supersonic flight and the flight characteristics of swept-wing supersonic aircraft. Flight tested at the Muroc Flight Test Facility alongside other research aircraft such as the X-1, X-1A, and X-2, the D-558-2 was the Navy's venture into the mysteries of supersonic flight. Controversy persists as to who deserves credit for the first Mach 2 flight. Crossfield reached Mach 2 in the D-558-2, but in a shallow dive. Just twenty-two days later, Chuck Yeager flew the Bell X-1A to Mach 2.44 in level flight.

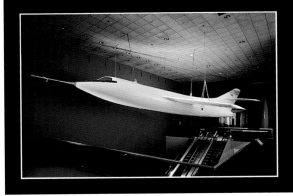

This beautiful, swept-wing airplane now hangs in the Milestones of Flight Gallery at the National Air and Space Museum.

On June 24, 1952, the NACA Committee on Aerodynamics called for an airplane that could probe the unknown problems of flight at Mach numbers between 4 and 10 and at altitudes between 12 and 50 miles. On October 5, 1954, this same committee, in executive session, made the final decision to proceed with this manned hypersonic research airplane, which would eventually become the X-15; Crossfield was a

member of the committee. On May 9, 1955, four aircraft companies submitted proposals to the Air Force (which was paying for the airplane): Bell, Douglas, North American, and Republic. After North American won the contract, Scott Crossfield left the NACA and joined North American as chief engineering test pilot and design consultant on the X-15.

After piloting the first test flight of the X-15 on June 8, 1959, Crossfield flew the airplane thirteen more times, his last X-15 flight taking place on December 6, 1960—the thirtieth test flight of the X-15 program. At this point, North American finished its contractor check flights and turned the aircraft over to the Air Force. Although Crossfield had expected to fly the X-15 during its entire program, because he was a NAA employee, not a NACA employee, his flight participation in the X-15 came to an end.

Crossfield continued with North American, first as the director responsible for systems tests, reliability engineering, and quality assurance for several aircraft and space vehicles, and then as its technical director, Research Engineering and Test. In 1967, he left the company to serve as a division vice president for Research and Development for Eastern Airlines until 1973, and he then served as senior vice president for Hawker Siddeley Aviation in 1974 and 1975. In 1977, nine years after the X-15 program ended, he became a technical consultant to the House Committee on Science and Technology. He served in this capacity for sixteen years, during which he was a steadfast proponent of manned hypersonic flight. He especially supported the massive U.S. X-30 supersonic combustion ramjet engine-(scramjet) powered single-stage to orbit aerospace plane project during the 1980s and early '90s. He retired in 1993.

Scott Crossfield earned a number of prestigious awards during his life, including being a joint recipient of the 1961 Collier Trophy, the

X-15 at rollout. *USAF, Air Force Flight Test Center History Office, Edwards Air Force Base*

International Clifford B. Harmon Trophy for 1960, the Lawrence Sperry Award for 1954, the Octave Chanute Award for 1954, and the Iven C. Kincheloe Award for 1960. He was inducted into the National Aviation Hall of Fame in 1983 and the International Space Hall of Fame in 1988. As a reflection on his aeronautical engineering accomplishments, the American Institute of Aeronautics and Astronautics elected him to the rank of Honorary Fellow in 1999, the highest recognition in that society. In 2000, the National Air and Space Museum awarded him its most prestigious award, the Lifetime Achievement Award. An elementary school in Herndon, Virginia, and the terminal of the Chehalis-Centralia Airport in Washington State both bear his name.

On April 19, 2006, Crossfield got into his Cessna 210A to return home from Maxwell Air Force Base in Montgomery, Alabama, where he had just finished giving a speech to a class of young Air Force officers. Amid severe thunderstorms, his airplane broke up in midair; recovery teams found wreckage in three different locations within a quarter-mile region. Later, the National Transportation Board ruled the probable cause of his crash to be a combination of two failures: Crossfield had not obtained updated weather information en route, and the air traffic controller failed to provide adverse-weather avoidance assistance. Crossfield was survived by his wife of sixty-three years, Alice Crossfield, as well as six children and nine grandchildren. He is buried in Arlington National Cemetery.

Engle in the X-15 cockpit with a view of the instrument panel. It's his first flight in the X-15 (October 7, 1963). *USAF, Air Force Flight Test Center History Office, Edwards Air Force Base*

Engle standing beside the X-15. *USAF, Air Force Flight Test Center History Office, Edwards Air Force Base*

Engle in the X-15 cockpit for his first flight. *USAF, Air Force Flight Test Center History Office, Edwards Air Force Base*

Pilot School, also at Edwards. The purpose of that school was to train military pilots to be astronauts.

Graduating in 1963, Engle was selected as a project pilot for the X-15 program. After his first X-15 flight in October, he went on to achieve Mach 5.71 on February 2, 1965. Typical of the maturing X-15 research program, on this flight Engle tested a Martin 255 ablative material on the ventral and nose panels, made skin friction measurements, checked out a nose gear modification, and took boundary layer noise data. On June 29, 1965, he reached 280,600 feet, qualifying him for an astronaut rating. His last flight in the X-15 was on October 14, 1965, which was also his third flight above an altitude of 50 miles.

In 1966, Engle was selected for the NASA astronaut program. He was thirty-two years of age, the youngest man to become an astronaut. He was also the only person in the program to have flown in space, by virtue of his X-15 experience. First assigned to the Apollo program, he was on the support crew for the Apollo 10 before becoming the backup lunar module pilot for Apollo 14. Since the Apollo program was coming to an end, he moved to the Space Shuttle program. In 1977, he was commander of one of the two crews that conducted atmospheric approach and landing tests with the Space Shuttle Enterprise. In November 1981, he commanded the second flight of the Space Shuttle Columbia (STS-2), during which he intentionally flew manually large portions of the reentry flight path, performing twenty-nine flight-test maneuvers from Mach 25 through landing. This was the first and only time a winged spacecraft has been manually flown from orbit to landing. His last flight into space was as commander of the Space Shuttle Discovery (STS-27) in August 1985.

Engle retired from the Air Force as a major general on November 30, 1986. He went on to participate in the Challenger disaster investigation in 1986 and consulted for the shuttle program

into the 1990s. He is enjoying his retirement as an aerospace and sporting goods consultant.

One of the more important aspects of the X-15 program was the providing of technical data for the design of the Space Shuttle. Joe Engel was the human link between the two programs, and he represents the rather smooth transition from the X-15 to the success of the Space Shuttle.

MILTON O. THOMPSON
1926–1993

Milt Thompson holds the distinction of being the only X-15 pilot to have written a book on the X-15 program. Entitled *At the Edge of Space: The X-15 Flight Program*, it was published by Smithsonian Institution Press in 1992, a year before Thompson's death. It is a highly recommended read for anybody interested in the inside story of the X-15 flight program. As the ninth test pilot to join the X-15 program, Thompson flew the airplane fourteen times, beginning on October 29, 1963. On November 27, 1963, the inertials failed at launch. On January 16, 1964, he reached Mach 4.92, but the speed brakes were extremely hard to open during the high aerodynamic heating phase. On February 19, at Mach 5.29, he had a premature burnout due to a clogged liquid oxygen line. His highest Mach number was 5.48, reached on January 13, 1965, during which he lost the pitch-and-roll damping mechanism during the pull-up/roll maneuver after burnout and temporarily lost control. His last flight in the X-15 was on August 25, 1965, when he achieved his highest altitude of 214,100 feet. The technical difficulties encountered by Thompson were typical of those encountered by all of the X-15 test pilots; there were very few totally "good flights" during the 199 flights of the airplane.

Milt Thompson was born on May 4, 1926, in Crookston, Minnesota. He became a naval aviator

Milt Thompson standing beside the X-15. *USAF, Air Force Flight Test Center History Office, Edwards Air Force Base*

LIFTING BODIES

A lifting body is a wingless aerodynamic configuration that generates its lift from the body at high angle of attack, somewhat like the Space Shuttle. In the period between the X-15 and the Space Shuttle, several "lifting bodies" were designed and flown to explore principally the subsonic characteristics of this hypersonic aerodynamic shape in order to provide data for the subsonic portion of the Space Shuttle flight.

at age nineteen and served in China and Japan during World War II. After six years of active duty, he left the Navy and entered the University of Washington, where he graduated with a bachelor's degree in aeronautical engineering in 1953. Following graduation, like many Washington graduates, he joined the Boeing Aircraft Company as a structural-test and flight-test engineer. He is one of only two X-15 pilots (along with Scott Crossfield) to have worked in the aircraft industry. One of the projects to which Boeing assigned him was testing the new B-52. In March 1956, he seized the opportunity to go to work for the

NACA's High Speed Flight Station at Edwards Air Force Base as a research pilot.

At the time, the NACA had only five pilots, including future X-15 pilots Joe Walker, Jack McKay, and Neil Armstrong. Thompson worked on the early X-airplanes. Of this experience, he admitted that he "watched apprehensively as these programs wound down and were terminated." He felt that the glory days of the X-airplanes were over and that he had missed it all. "In the next few years," he later wrote, "I realized that I was wrong. The golden years were still to come."

For Thompson, those glory years began when he was selected by the Air Force to be the only civilian pilot on the X-20 Dyna-Soar winged hypersonic vehicle project. Although he again witnessed yet another cancelation when the Dyna-Soar project was prematurely stopped, his participation on lifting entry bodies continued. He was the first person to fly such a lifting body, the lightweight M2-F1. He continued to fly this aircraft a total of forty-seven times, after which he made the first five flights in the all-metal M2-F2. He took all this experience to the X-15 program.

Thompson finished his active flying career in 1967. Two years later, he became chief of Research Projects, and in 1975 he was appointed chief engineer, a position he held until his death on August 6, 1993.

WILLIAM J. KNIGHT
1929–2004

During the course of his sixteen flights in the X-15, William "Pete" Knight experienced perhaps the most notable event of all the pilots who flew the airplane. On October 3, 1967, he achieved Mach 6.7, the fastest speed attained in the X-15.

Pete Knight kneeling beside the X-15. *USAF, Air Force Flight Test Center History Office, Edwards Air Force Base*

Knight standing beside the X-15 after a flight. *USAF, Air Force Flight Test Center History Office, Edwards Air Force Base*

Knight in the cockpit of the X-15 after a flight. *USAF, Air Force Flight Test Center History Office, Edwards Air Force Base*

By virtue of this flight, Pete Knight still holds today the world's speed record in a winged, powered aircraft.

On this same flight, the X-15 was coated with a white ablative heat shield. Attached underneath the X-15 was a dummy model of NASA's high-speed research engine (HRE), part of a research program to develop a supersonic combustion ramjet engine (scramjet). During the course of the test, the shock wave from the engine cowling impinged on the bottom surface of the X-15. The intense aerodynamic heating in the impingement region burned through the attachment pylon, separating the dummy scramjet from the airplane. Had the dummy engine remained attached any longer to the airplane, the shock wave would have burned a hole into the primary structure of the fuselage

and most likely would have resulted in destruction of the X-15 in flight. Moreover, this was the last flight of the X-15A-2. The airplane is now on permanent display in the Air Force Museum at Wright-Patterson Air Force Base in Ohio.

Pete Knight was born on November 18, 1929, in Noblesville, Indiana. At the age of twenty-one, he enlisted in the Air Force, and he obtained his pilot's wings in 1953. He was assigned to the 438th Fighter-Interceptor Squadron, flying Northrop F-89 Scorpions. While flying the F-89, he entered the National Air Show at Dayton, Ohio, in 1954 and won the prestigious Allison Jet Trophy, becoming one of the youngest pilots to win the award. He then began his engineering study program, and he graduated from the Air Force Institute of Technology in 1958 with a

Knight examining the X-15-2 covered with the white ablator coating. *USAF, Air Force Flight Test Center History Office, Edwards Air Force Base*

bachelor's degree in aeronautical engineering. With his career on a fast track, he graduated from the Air Force Test Pilot School that same year. Assigned to Edwards Air Force Base, he was a project test pilot for the F-100, F-101 Voodoo, F-104 Starfighter, T-38, and F-5.

The Air Force recognized Knight's expert piloting ability by selecting him in 1960 to be one of the six test pilots for the X-20 Dyna-Soar, a winged orbital space vehicle that was an early precursor to the Space Shuttle. The X-20 program was canceled in 1963, but Knight went ahead to complete the Air Force astronaut training program at Edwards Air Force Base. With this background, Pete Knight became the tenth X-15 test pilot, and he had his first flight in the airplane on September 30, 1965. He flew the X-15 sixteen times. On October 17, 1967, he achieved an altitude of 280,500 feet, qualifying him for official astronaut status.

On June 29, 1967, Knight experienced total power failure while going through 107,000 feet at Mach 4.17. All onboard systems shut down.

He coasted to a maximum altitude of 173,000 feet and calmly set up a visual landing approach. He resorted to the old "seat-of-the-pants" flying and glided safely to an emergency landing at Mud Lake, Nevada. For this expert example of flying, he earned a Distinguished Flying Cross. On July 16, 1968, Knight had a hydraulic gauge malfunction during boost, which required him to push over to an alternate flight profile, which is the planned variation of speed, altitude, and location for the flight of the aircraft. On his glide back to Edwards, he experienced unexpected shaking and vibrations. His last flight in the X-15 was on September 13, 1968; this was the 198th flight of X-15, the next to last flight of the program.

Pete Knight went on to a stellar Air Force career. He went to Southeast Asia in 1969 and completed a total of 253 combat flights in the F-100. His testing career was then extended to the F-15 program at Wright-Patterson Air Force Base as test director; he became the tenth pilot to fly the F-15 Eagle. He then returned to Edwards in 1979 as vice commander of the Air Force Flight Test Center. After thirty-two years of service and more than 6,000 hours in the cockpits of more than a hundred different aircraft, he retired from the Air Force as a colonel in 1982.

Knight became the only X-15 pilot to go into politics. In 1984, he was elected to the city council of Palmdale, California, and he became the city's first elected mayor four years later. After becoming the fastest airplane pilot in the world, he thus became mayor of the fastest growing city in the United States. He was elected to the California State Assembly in 1992 and to the California State Senate in 1996. Knight achieved widespread public notice as the author of Proposition 22, the purpose of which was to ban same-sex marriage. He continued to serve in the California State Senate, representing the 17th District, until his death on May 7, 2004.

WILLIAM HARVEY "BILL" DANA
1930–

Bill Dana was the eleventh X-15 test pilot. He flew the X-15 sixteen times and was the pilot for the 199th flight, the last of the X-15 program.

Bill Dana was born in Pasadena, California, on November 3, 1930. He attended the United States Military Academy at West Point, graduating with a bachelor of science degree in 1952. He satisfied his military commitment by serving as a pilot in the U.S. Air Force for four years, after which he attended the University of Southern California. At USC, he graduated with a master of science degree in aeronautical engineering in 1958. He began his distinguished civilian career at the Dryden Flight Research Center on October 1, 1958.

This was the first day that NASA went into operation, and Dana proudly became NASA's first employee. He was involved with the X-15 from that first day, initially as an engineer, then as a chase pilot, and finally as a project pilot. His first X-15 flight was on November 4, 1965, a checkout flight during which he reached Mach 4.22 and an altitude of 80,200 feet. At this point in the X-15 program, even the pilot checkout flights were relatively high-performance. This flight required two relights of the rocket engine. On October 4, 1967, Dana reached his highest speed, Mach 5.53, and on November 1, 1966, he achieved his highest altitude of 306,900 feet, one of two flights he made above 50 miles.

By the end of the X-15 program, Dana was just at the beginning of his distinguished career as a test pilot and aeronautical engineer. Building on his experience flying the X-15, he became a project pilot for NASA's manned lifting body program, a precursor to the Space Shuttle. He completed one NASA M2-F1, nine Northrop HL-10, nineteen Northrop M2-F3, and two Martin Marietta X-24B flights, for a total of thirty-one lifting body missions. For this work, he received the NASA Exceptional Service Medal.

In 1976, Dana received the Haley Space Flight Award from the American Institute of Aeronautics and Astronautics. In 1986, he became the chief pilot at the Flight Research Center, and he then became the assistant chief of the Flight Operations Directorate. He continued to fly on several important research programs: the F-15 Highly Integrated Digital Electronic Control and the F-18 High Angle of Attack program. In August 1993, Dana became chief engineer of the NASA Dryden Flight Research Center, and he held that position until his retirement in 1998.

After retirement, Dana began a distinguished second career by working as a contractor with the NASA Dryden History Office. He was honored by the Smithsonian's National Air and Space Museum in 1998 when he was selected to give the Charles A. Lindbergh Memorial Lecture, the most prestigious lecture at the museum. His lecture title was "A History of the X-15." He still continues to lecture and write papers based on his experience in high-speed flight.

MICHAEL J. ADAMS
1930–1967

Mike Adams was the twelfth (and last) pilot in the program, and he was the only pilot to lose his life flying the X-15.

On November 15, 1967, Michael Adams, veteran pilot with six previous X-15 flights, entered the aircraft for a flight to evaluate a guidance display and to conduct several experiments. He had spent more than 21 hours practicing the specifics of this flight in the simulator. The drop at about 10 a.m. and 45,000 feet was normal, and he climbed to 266,000 feet. While the aircraft climbed to higher altitude after launch, an electrical disturbance caused the MH-96 dampers to trip out. Adams reset the dampers. He then switched the sideslip indicator to a vernier

Mike Adams in the X-15 cockpit before his first flight, October 6, 1966. *USAF, Air Force Flight Test Center History Office, Edwards Air Force Base*

attitude control mode to more accurately control the experiments. He planned to reset this back to indicate yaw angle when returning to base in order to see his sideslip during approach to landing. But this instrument change prevented him from seeing that the airplane was yawing at a critical time in the flight.

After burnout, as he soared upward, he conducted a wing-rocking experiment, in which the rocking became excessive as he approached his peak altitude, 266,000 feet. His yaw had drifted to 15 degrees, and he was unaware of this because his instrument was inadvertently set to show pitch attitude, not yaw. About 15 seconds later, the airplane was yawing wildly and Adams

communicated to Pete Knight that "the airplane seems squirrelly." He soon after stated that he was in a spin, subjected to high accelerations. Since little was known about the hypersonic spin characteristics of the airplane, the ground crew was not able to offer advice. According to the ground data that was later correlated with the flight data, when Adams recovered, he was yawed 90 degrees, flying upside down, and descending at supersonic speed.

Adams pulled out of the spin, and he probably would have had a successful landing except that the MH-96, the Minneapolis-Honeywell adaptive flight control system, was on and locked in, causing the airplane to oscillate between its limits,

Adams suited up and walking to the X-15 for his first flight. *USAF, Air Force Flight Test Center History Office, Edwards Air Force Base*

up and down, preventing Adams from correcting his attitude and flying his way home. The loads on the airplane built up beyond the structural limits, and the X-15-3 aircraft broke up at approximately 62,000 feet and about 3,800 feet-per-second speed. It crashed to the desert floor near Johannesburg, California. There was talk about Adams having slight vertigo, which may have contributed to his not noticing the yaw buildup or resetting the yaw indicator to the yaw setting.

Adams's death shows the dangers of flight testing a new aircraft in previously untested regions of flight, and of flying experiments in which certain research-data measuring instruments may have caused an electrical disturbance that affected the MH-96 from operating at its top quality and in conditions it was not designed for. Any and all these things may have influenced the accident.

Because his flight was above 50 miles high, Adams was posthumously awarded an astronaut rating. For the X-15 program, the tragedy was a blight, but it was the only casualty in 199 flights. Since the objectives for the airplane had been accomplished, the accident was a major reason for the termination of the X-15 program. There were only seven subsequent flights.

Michael Adams was born on May 5, 1930, in Sacramento, California. After graduating from Sacramento Junior College, he enlisted in the Air

ROTATIONAL MOTION

The rotational motion of an airplane in flight takes place centered around the airplane's center of gravity. It is a combination of three rotational directions: the nose up or down rotation, called pitch; the wing rotation about the fuselage, called roll; and the nose swinging right or left, called yaw.

Force in November 1950. The Korean War was in full force at that time, and Adams flew forty-nine combat missions as a fighter-bomber pilot in Korea. In 1958, he earned an aeronautical engineering degree from the University of Oklahoma, and he went on to eighteen months of study at MIT in astronautics. In 1962, he was selected to attend the Experimental Test Pilot School at Edwards Air Force Base. He excelled at the school, winning the Honts Trophy as the best scholar and pilot in his class. In December 1963, he graduated with honors from the Aerospace Research Pilot School. His first flight in the X-15 was on October 6, 1966. On June 8, 2004, a memorial monument to Adams was erected near the crash site, northwest of Randsburg, California.

Test pilots are a special breed. They face risks above and beyond those faced by conventional pilots. The X-15 pilots, however, are in a special class. They were *research* test pilots, putting their lives on the line to prove the viability of a pioneering hypersonic airplane and to obtain research data on an unknown regime of flight. This data was invaluable to the subsequent design of the Space Shuttle.

On almost every flight of the X-15, some type of technical problem or failure occurred, sometimes multiple problems on the same flight.

Signed photo of six of the X-15 pilots standing beside the X-15. From left to right: Rushworth, McKay, Peterson, Walker, Armstrong, and White. *USAF, Air Force Flight Test Center History Office, Edwards Air Force Base*

X-15 mounted under the wing of the B-52 prior to a flight. *USAF, Air Force Flight Test Center History Office, Edwards Air Force Base*

X-15 on the lakebed of Rogers Dry Lake. *USAF, Air Force Flight Test Center History Office, Edwards Air Force Base*

It is remarkable that only one pilot, Mike Adams, lost his life during the whole X-15 program of 199 flights. Ten of the twelve had formal college degrees in aeronautical engineering and took pride in their status as dedicated, professional aeronautical engineers. All served at one time or another in the military, and six (Crossfield, Walker, McKay, Armstrong, Thompson, and Dana) were in civilian status when they flew the X-15. Of the career military officers who flew the X-15, three retired as major generals in the Air Force and one as a vice admiral in the Navy.

B-52 flying over the X-15 on the ground. *USAF, Air Force Flight Test Center History Office, Edwards Air Force Base*

06 | THE FLIGHTS

RISKS IN RESEARCH AIRCRAFT
When we talk about risk, we mostly mean the life of the pilot, the dangers to the man who governs the airplane through its flight path to the new conditions in flight that the new airplane will investigate. This is the life of a person who is talented, productive, and well experienced in test flying—and a human being unique in his flying abilities in high-speed and high-altitude flight. These characteristics are in addition to all the other attributes that pertain to each person's life. We also mean the risk to the airplane, which is important enough to have had many years of development, thousands of man-hours of workmanship, and millions of dollars in cost. If the airplane is lost, the research program for which it was designed is jeopardized.

Bob White
in the X-15
Force Flig
History Of
Air Force

X-15 at the end of Jack McKay's flight on May 6, 1966, during which the rocket engine failed after 35.4 seconds. The X-15 landed at Delamar and skidded off the smooth lakebed. McKay was not injured, and the X-15 sustained only slight damage. *USAF, Air Force Flight Test Center History Office, Edwards Air Force Base*

This research aircraft was designed to explore regions for flight at altitudes and speeds not yet then achieved. It incorporated a new, advanced design using new materials that allowed it to operate at higher temperatures than previously experienced. It had new and multiple control systems. It used a new rocket engine with a new fuel-oxidizer combination. It had twin skids rather than wheels for a landing gear. Wind tunnel data existed for the aerodynamics in this new speed region, but it had not been evaluated and confirmed in flight. Moreover, not all the conditions of hypersonic flight that it experienced in the wind tunnel had been previously analyzed or fully understood. Some problems were not known until they were discovered in flight. Therefore, they could not have been addressed in advance.

Before this aircraft could achieve hypersonic speeds and high altitudes, it still had to traverse all the flight regions previously explored. The new design had to prove that it could safely fly in those known flight regions. For example, it had to be able to take off on its own or be air-dropped in the subsonic regime. It then need to accelerate to high subsonic speeds, go through transonic flight, experience shock waves beginning at Mach 1, and accelerate to supersonic speeds, experiencing stability changes longitudinally, and thereafter in regions of reduced lateral-directional stability with increasing Mach number. It also had to decelerate and return to the landing site with normal approach, descent, and landing, all without using thrusting power. It should be noted that although low-speed subsonic flight and landing had been analyzed for the X-15 for these conditions, they were not the primary focus of the design.

The pilots controlled many aspects of the flight, such as the handling and control actions about the three axes of the airplane and the application of thrust. But the pilots could not control other factors, such as the strength of materials at high temperatures and the effect of temperature gradients on the design and strength caused by high aerodynamic heating on the outside and cool internal temperatures.

The characteristics of the X-15 would not be definitively known and understood until verified or determined in actual flight. The handling characteristics in these regions were unique, controlled by the pilot with three different control systems: a traditional stick on the floor between the legs and a rudder; a small control stick on the right console, with power assist or electronic force amplification when experiencing dynamic pressures too high for normal pilot forces; and a rocket power control on the left console for use in space where the air is too thin and the dynamic pressure too low for aerodynamic control surfaces to be effective. In the X-15, the pilot experienced for the first time these new controls, designed for this airplane, following his drop from the B-52 at altitudes of about 40,000 feet and speeds of about Mach 0.8. There were no ground trials with the controls during taxiing or on short hops prior to a real test flight, as is possible while familiarizing oneself with the controls of a conventional aircraft that has wheels rather than skids and that has a jet or reciprocating engine instead of a rocket.

Pilots had to address new interfaces in each new test aircraft. For example, the X-15 was taken aloft by the B-52 and attached under the

Discussion before a flight. *USAF, Air Force Flight Test Center History Office, Edwards Air Force Base*

B-52's right wing, unlike the other rocket research aircraft. These previous X-airplanes were attached under the fuselage, allowing the test pilot to ride in the mother craft's cabin and enter the test aircraft only after everything had been checked out. In the X-15, located out on the wing, the pilot had to enter his aircraft before B-52 takeoff, and he was at risk as the two airplanes climbed to altitude. He had to also check out the X-15 systems while riding in the X-15 after takeoff and prior to drop. He thus had to deal with the interface with the mother airplane mechanically and electronically, including communications, and also operationally by topping off the liquid oxygen and checking other conditions before separation and drop occurred in midair. In

Scott Crossfield in the cockpit of the X-15. *USAF, Air Force Flight Test Center History Office, Edwards Air Force Base*

Crossfield in discussion after a flight. *USAF, Air Force Flight Test Center History Office, Edwards Air Force Base*

both the X-15 and B-52, the interfaces included the mechanical ties between subsystems and components, each of which had their own requirements, as well as continuity of electrical and electronic signals that signaled to and from the pilot's cabin and the operating systems, as well as the thermal interfaces that controlled the heating and cooling needs of particular subsystems and components. Repairs could have adversely affected these initial design conditions. An example of this occurred when extra cooling was needed for a component after problems in flight and an extra cooling line was installed to fix the problem. This new cooling line ran alongside an APU hydraulic line that caused its hydraulic fluid to freeze, preventing the APU from functioning.

A conventional aircraft has a shakedown period in which the newly installed subsystems first operate together as a complete aircraft system. Then interfaces with other elements of the airplane are tested mechanically, electrically, and thermally in actual flight conditions when they can be fine-tuned. Experience has shown that many changes or improvements are necessary in a new airplane. Routinely, there are bugs to work out, safety issues to resolve, and procedures to establish. For the flight-test program, there is a new support team from management through inspection. For an airliner, it may take two years of testing before it is put into use. There is no such luxury for these high-performance research aircraft. They start out in their very first flight at 40,000 feet in the air. Experience also has shown that unexpected difficulties are uncovered in air-launched research aircraft such as the X-1, X-1A, and X-2, in the increasing velocity regions of transonic and supersonic flight.

Such a flight program is necessarily risky. This was a new airplane. The old flight regimes in which this plane had to traverse were not the prime focus in design. New equipment, previously untested in flight, was necessary, and the exploration was conducted in a new flight regime to ascertain the validity and shortcomings of the applicable theories, which were approximated with many simplifying assumptions and the use of wind tunnel test data.

Since the X-15 followed the course of the previous X-aircraft, it also had numerous difficulties with equipment—such as the auxiliary power units, landing gear, windshield and cockpit seals, stability in landing, and so forth—that required pilot experience, fortitude, and ingenuity to overcome. In the 199 flights, the problems were frequent, unanticipated, and in many instances life-threatening. It was the piloting excellence, the prior experience of the pilots and engineers, and the extensive preparation for each flight—including hours of simulation—that permitted these many flights to be completed with only one fatality.

Another difference from conventional aircraft testing relates to the lack of any power when the rocket fuel is expended. The fuel is used up in just about 90 seconds of flight. Conventionally

DAVID CLARK PRESSURE SUIT

The David Clark full-pressure suit was developed by Dr. David Clark and produced in his small factory in Worcester, Massachusetts. Unlike previous partial-pressure suits that pressurized only parts of the human body, Clark's full-pressure suit provided pressurization for the whole body. It was made from his patented Link-Net nylon fabric, which consisted of two layers of nylon arranged with opposite bias that provided maximum strength in high-stress areas while also allowing the suit to deform easily to the pilot's movement. It was lightweight, but it held its shape under pressure. The suits were custom-made for each pilot, who had to make several trips to Worchester for fitting. Clark made improvements to the suit throughout the X-15 program. It became the standard full-pressure suit for the Air Force and NASA, being used by pilots of the U-2 and SR-71 high-altitude spy planes as well as the Space Shuttle astronauts. Several photographs in Chapter 5 show some X-15 pilots in their David Clark full-pressure suits.

powered aircraft can reposition themselves if in trouble or when in descent, approach to landing, and during the landing itself. All X-15 flight positions and corrections have to be done with aerodynamic controls alone, not with power. If the landing approach is too high or too low, the pilot must bring it down safely without power. He cannot go around the field a second time to try again. His first attempt must be successful.

Joe Walker being congratulated after completing his flight on August 22, 1963, during which he set the highest-altitude record of 354,200 feet. *USAF, Air Force Flight Test Center History Office, Edwards Air Force Base*

HIGHLIGHTS OF THE FLIGHTS

What is it like for a research test pilot to fly an X-15 airplane into unknown areas of speed and altitude? He arrives early in the morning, a good time for flight since the winds and temperature are lower at that time in this desert area. He goes to the physiological van at Edwards Air Force Base and there puts on his David Clark full-pressure suit. He walks across the ramp to the airplanes, the B-52 and X-15. He climbs a large ladder to a platform next to the X-15, and then he enters the small X-15 cockpit. He prepares the airplane and himself for takeoff while the X-15 is attached to the B-52 mother plane.

The B-52 crew goes through a preflight list that includes the location, altitude, and velocity at which the X-15 was to be launched. They then start the engines and check that everything is okay with the pilot, who is captive in the X-15 under the wing. (All this was a much less severe routine than that required by the X-15 pilot in preparation for the flight, but their job to make sure the X-15 was safely launched was just as important.)

The B-52 takes off and climbs to altitude, about 45,000 feet. There the flight crew inside the B-52 prepares for the drop launch of the X-15, going through their checklist and topping off the liquid oxygen in the X-15, some of which has boiled off during the climb to launch altitude.

When all is ready, the B-52 drops the X-15, located underneath its right wing. The X-15 smoothly separates from the mother ship, usually with a roll to the right to compensate for the local airflow located under the right wing of the B-52. The X-15 pilot levels his airplane and lights up his engine. He accelerates away from the B-52 and, once clear, the pilot rotates his airplane to increase the angle of attack for climb to altitude.

Although the primary purpose of the X-15 was the acquisition of research data on the aerodynamics, thermodynamics, and flight dynamics of hypersonic flight, the quest for speed and altitude has been the driving force in the historical advancement of the airplane over the past 120 years. Therefore, obtaining maximum speed and maximum altitude was also important. However, the flight conditions required to obtain maximum speed are different than those to obtain maximum altitude.

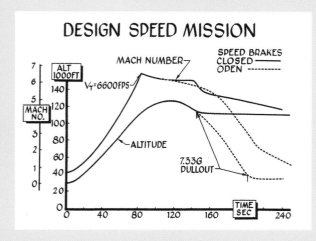

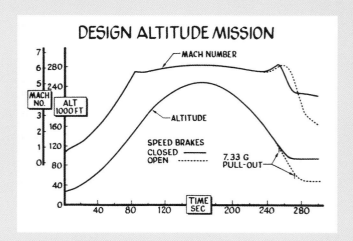

MAXIMUM SPEED

Here, the pilot continues his climb to altitude, then pushes over at zero lift until the airplane is in level flight at the desired altitude. He continues to fly at that altitude at full thrust until the maximum speed is obtained, which occurs when the fuel is used up. Zero lift means that the pilot adjusts the orientation of the airplane relative to the airflow ahead of the airplane (the angle of attack) so that the aerodynamic lift becomes zero, and he holds this until the X-15 is now moving in horizontal flight (level flight).

The airplane then starts to fall back to earth under the force of gravity, and it decelerates as the aerodynamic drag builds up at lower altitudes. During this return to earth, the airplane is in a steep glide, with a plan to reach an altitude of about 35,000 feet with a velocity of 290 to 350 miles per hour (called high key, which was the highest approach to the runway at Edwards Air Force Base). From there, he descends to an altitude of 18,000 feet, flying in the opposite direction of the landing runway (called low key on the flight trajectory). At this point, the airplane is about 4 miles from touchdown. The pilot continues in a 180-degree turn and then lands, probably at a speed of 200 miles per hour.

MAXIMUM ALTITUDE

After launch from the B-52, the X-15 continues to climb until the fuel is used up and then continues in an upward ballistic trajectory, reaching a maximum altitude determined by its kinetic energy at the point of engine burnout and the force of gravity. The airplane then begins to descend. The pilot then heads for home, reaches high key above Edward, descends, and lands as above. Because of the high altitude, the glide return is over a larger distance than the lower-altitude flights. For these flights, the airplane would be dropped at a greater distance from Edwards Air Force Base, sometimes as far as away as 300 miles, so that his glide ends at Edwards.

For most of the X-15 flights, the data gathering was done in the regions bounded by the maximum speed and the maximum altitude flights. The variation of Mach number and altitude during these flights is shown in the two Mach number/altitude versus time-of-flight figures shown, one for a maximum speed flight and one for a maximum altitude flight.

The data obtained in the hypersonic region of these flights provided vital flight data points that were calibrated against analytical predictions and against wind tunnel data. The designing of aircraft

Arrival of the first X-15 to Edwards Air Force Base. *USAF, Air Force Flight Test Center History Office, Edwards Air Force Base*

Unloading the X-15 upon arrival at Edwards Air Force Base. *USAF, Air Force Flight Test Center History Office, Edwards Air Force Base*

The welcoming crowd upon arrival of the X-15 to Edwards Air Force Base. *USAF, Air Force Flight Test Center History Office, Edwards Air Force Base*

PREPARATION FOR FLIGHT—UNDER THE WING OF THE B-52

X-15 being mated to the B-52. *USAF, Air Force Flight Test Center History Office, Edwards Air Force Base*

X-15 mated to the B-52. *USAF, Air Force Flight Test Center History Office, Edwards Air Force Base*

TAKEOFF

Detail of the mating of the X-15 with the B-52 for its first flight with external fuel tanks (empty), November 3, 1965. *USAF, Air Force Flight Test Center History Office, Edwards Air Force Base*

Takeoff of the B-52 with the X-15 with external tanks, November 3, 1965. *USAF, Air Force Flight Test Center History Office, Edwards Air Force Base*

MATED WITH B-52 IN FLIGHT

X-15 mated with the B-52 for one of its early contractor flights. *USAF, Air Force Flight Test Center History Office, Edwards Air Force Base*

Takeoff of the B-52 with the X-15 mounted under the wing. *USAF, Air Force Flight Test Center History Office, Edwards Air Force Base*

LAUNCH

X-15 mounted under the wing of the B-52 mother ship at altitude. *USAF, Air Force Flight Test Center History Office, Edwards Air Force Base*

X-15 at the moment of engine ignition and launch from the B-52. *USAF, Air Force Flight Test Center History Office, Edwards Air Force Base*

X-15 in flight after launch. *USAF, Air Force Flight Test Center History Office, Edwards Air Force Base*

X-15 in flight. *USAF, Air Force Flight Test Center History Office, Edwards Air Force Base*

LANDING ON LAKEBED AND RESTING ON LAKEBED

X-15 landing with the F-104 chase plane alongside. *USAF, Air Force Flight Test Center History Office, Edwards Air Force Base*

X-15 after landing. *USAF, Air Force Flight Test Center History Office, Edwards Air Force Base*

X-15 resting on Rogers Dry Lake after a flight. *USAF, Air Force Flight Test Center History Office, Edwards Air Force Base*

X-15-3 on the lakebed. *USAF, Air Force Flight Test Center History Office, Edwards Air Force Base*

X–15A-2 with external fuel tanks on the ramp of the NASA Flight Research Center at Edwards. *USAF, Air Force Flight Test Center History Office, Edwards Air Force Base*

to fly in these regions, as well as vehicles to return from space, could proceed with confidence by knowing what corrections to make to the analyses and wind tunnel data. This data gathering and its correlation to analysis and wind tunnel results was the purpose of the X-15 research airplane program.

On October 3, 1967, Pete Knight achieved the maximum Mach number for the X-15, and he did it flying the modified version of the X-15, the X-15-A2, with additional fuel in the extended fuel tanks and with extra external fuel tanks. The extra fuel allowed more full thrust time, totaling 141 seconds—50 seconds more than the basic X-15 Nos. 1 and 2. After being in the X-15 for more than an hour under the wing of the B-52 while on the ground, Knight performed the preflight checklist and was lifted when the B-52 took off at 1:20 p.m. They headed for Mud Lake, over which the B-52 dropped him an hour later.

It took two launch attempts before the drop actually worked. Knight stated later that he "reached up and hit the launch switch and immediately took my hand off to [go] back to the throttle and found that I had not gone anywhere. It did not launch." [citation: Jenkins, *X-15: Extending the Frontiers of Flight*, NASA SP-2007-562, 1967, p. 459] A second attempt 2 minutes later resulted in a smooth release. Pete then accelerated and climbed at an angle of attack of 12 degrees (angle between the wing chord and the free-stream airflow direction) at high lift until he reached a climb angle (angle between the horizontal and the flight path) of 32 degrees. He leveled off at 102,100 feet and reached a speed of 6,600 feet per second (Mach 6.7). This speed remains the fastest for a manned-powered airplane forty-seven years later, with no competitor airplane in sight.

Then, some unpleasant excitement occurred after burnout. Pete performed some rudder pulses to get data with the yaw damper off. As he decelerated through M=5.5, the "Hot Peroxide" warning light came on. On this particular flight, the X-15 was carrying a dummy supersonic combustion ramjet engine (scramjet) below its fuselage as part of a NASA hypersonic propulsion project. This was not an operating engine; it was a dummy engine being carried under the X-15 to examine the aerodynamic characteristics of the engine shape in full-scale hypersonic flight. The warning was caused by the aerodynamic heating generated by the shock wave from the dummy scramjet impinging on the bottom surface of the X-15. It severely damaged the airplane. Pete jettisoned the remaining peroxide to prevent it from exploding. The dummy scramjet was externally mounted in anticipation of future experiments. Shock waves also impinged on the vertical tail, with some melting and skin rollback.

The hot-peroxide event distracted Knight from energy management of the X-15, and he arrived at high key at supersonic speed rather than the desired, slower, subsonic speed. With this airspeed, the X-15 had too much kinetic energy. Pete then tried to jettison the ramjet, but nothing seemed to happen. He dissipated the excess kinetic energy by flying past the landing site, allowing aerodynamic drag to slow the airplane, and then landed at the proper speed. The dummy ramjet didn't release at once when jettisoned, and it was later located on the lakebed after some clever reasoning and analysis by Johnny Armstrong of the Flight Planning Group.

Joe Walker flew the maximum altitude flight on August 22, 1963. In his prior flight on July 19, 1963, the maximum altitude planned by NASA for that flight had been 315,000 feet, but he unintentionally overshot that mark and achieved an altitude of 347,800 feet, close to the maximum altitude of 360,000 feet that NASA was ultimately seeking for the X-15. The airplane could go over 400,000 feet, but there was concern about the reentry from that altitude. It was deemed difficult but possible for the pilot to make a successful reentry from there, but NASA set a limit at 400,000 feet. Because of the risks of reentry from higher altitudes, they set the flight at 360,000 feet to allow for the inaccuracies of the engine and the ability of the pilot to hold to the tight limits of controlling the angle of attack.

The flight path was selected, with climb angles and fuel cut-off that were calculated to achieve their goal. The engine thrust could vary from 57,000 pounds to 60,000 pounds, and a difference of 1,500 pounds would result in a 7,500-feet altitude change. One second in fuel cut-off time would result in a 4,000-foot altitude change, and if the climb angle were off by one degree, a 7,500-foot change in altitude would result. The planned maximum altitude of the flight was set at 360,000 feet because it allowed a factor of safety. If some of the slight variations in engine thrust, fuel cut-off time, and climb angle took place, the inadvertent increase in altitude would not take the X-15 to over 400,000, where reentry was more dangerous.

This flight was delayed for about two weeks because of weather and airplane APU problems. The actual launch went well, and Walker stayed close to the flight plan. The propellants were depleted at 176,000 feet at a speed of 5,600 feet per second. The airplane continued to soar upward on a ballistic trajectory to 354,200 feet—two minutes after fuel burnout. At that point, Walker and the X-15 were 67 miles high.

After reaching peak altitude, the airplane headed home, some 306 miles away, and was moving at 5,500 feet per second when it passed through 176,000 feet. This was a mirror image of

Joe Walker, X-15 test pilot (left), and John McTigue, NASA project engineer for the X-15-3 (right), in front of the X-15 mated to the B-52. *USAF, Air Force Flight Test Center History Office, Edwards Air Force Base*

ASTRONAUT WINGS

The Air Force pilots who flew the X-15 to altitudes above 50 miles all received Astronaut Wings, but NASA had decided not to give the same award to the civilian pilots who had made the same achievement. This caused controversy within the aerospace community. Finally, NASA reversed this policy, and in a ceremony on August 23, 2005, the three NASA pilots who flew the X-15 above 50 miles—William Dana, Jack McKay, and Joe Walker—were awarded Astronaut Wings. Only Bill Dana was alive at that time to receive the certificate. However, the families of McKay and Walker were present to receive the honor.

its ballistic climb after fuel burnout. The pullout force at 5 g occurred at 95,000 feet, and the pilot maintained the high g pullout in order to level flight at 70,000 feet. The rest of the flight back to landing at Edwards Air Force Base was uneventful. The total time of flight was 11 minutes and 8 seconds. While 67 miles is well above the 50 miles required for the pilot to achieve official astronaut rating, it was not awarded to Joe Walker until forty-two years later, after he had died.

There was only one fatal accident during the whole X-15 flight-test program. On November 15, 1967, Michael Adams lost his life when a possible electrical disturbance affected his flight control

Pete Knight and the X-15. *USAF, Air Force Flight Test Center History Office, Edwards Air Force Base*

Mike Adams in the cockpit of the X-15 (mated to the B-52), in preparation for his first X-15 flight, October 6, 1966. *USAF, Air Force Flight Test Center History Office, Edwards Air Force Base*

The X-15A-2 with its ablation coating. *USAF, Air Force Flight Test Center History Office, Edwards Air Force Base*

system. This, combined with his possible vertigo, caused his X-15 to go out of control and break up at an altitude of approximately 62,000 feet during descent and crash to the desert floor. This flight underscored the risk involved in such flight testing. The details of this flight are given in Chapter 5.

NUMBER OF FLIGHTS CONDUCTED

In all, 199 flights were conducted over a nine-year period from June 1959 to October 1968. Three airplanes were built, repaired, and rebuilt during that period. The third airplane was a significant modification. This longer version included external fuel tanks to extend the flight time, the range of altitude, and the Mach number to be investigated. Most of the initial objectives for the airplane were reached in the early years. But because the X-15 could fly in the hypersonic regime, NASA wanted to conduct many experiments, some examining various materials using the airplane as a test bed.

ABLATIVE COATING

One of the thermal protection techniques used to protect hypersonic vehicles from the intense aerodynamic heating environment is the covering of the vehicle surface with an ablative material. This material would directly absorb the heat and burn away (ablate), thus protecting the surface underneath. Some of the later X-15 test flights tested a specific ablative material, namely MA-25S developed by Martin Marietta. This silicon-based material was sprayed on the surface of the X-15. After several hours of curing, it was sprayed with a coating of Dow Corning DC90-090, a silicon-based sealer, which gave the X-15 a white color.

The MA-25S ablator coating used on the X-15A-2 had a pink color, as shown here. It was then covered with a protective layer of white Dow DC90-090 sealer, giving the X-15 a white color for the ablator experiments. *USAF, Air Force Flight Test Center History Office, Edwards Air Force Base*

Some of these caused problems in flight. For example, for some flights an ablative material was put on the airplane for testing purposes and for additional heat protection. As the material vaporized, it coalesced on the windshield, making it opaque, seriously affecting the visibility of the pilot. For further tests of the ablating material, the engineers had to install an external shield on half the windshield that could be moved away after ablation had obscured the other side in order to allow the pilot to have clear vision for the remainder of the flight.

LATERAL DIRECTIONAL STABILITY AND THE MH-96 FLIGHT CONTROL SYSTEM

As expected on the basis of experience with the earlier supersonic X-airplanes, the lateral-directional stability of the X-15 decreased as the Mach number rose to supersonic and hypersonic speeds. Honeywell's adaptive control system automatically compensated for the aircraft's unstable lateral-directional behavior in various flight regimes, and it utilized the combined operation of the aerodynamic control surfaces and the rocket reaction controls in their respective regions of flight.

Originally, the vertical tail sections above and below the airplane were large. That section, located below the airplane, is called the ventral tail. Wind tunnel data showed a need for a large ventral tail, so large that it would hit the ground first before the landing skids. This necessitated designing the bottom part of the ventral to be ejected prior to landing. The flight data showed a lesser need for the large area of the ventral tail, and in subsequent flights the bottom half was left off.

A relationship between the wind tunnel data and the flight data was thus established. The Honeywell MH-96 adaptive control system allowed the airplane, unstable in certain regions of flight, to be operated in a conventional manner throughout. Moreover, it provided an automatic transition from the conventional aerodynamic control system (rudder, elevator, etc.) used within the sensible atmosphere to the reaction control system for high-altitude flight, where the aerodynamic forces were too weak. This relieved the pilot from manually making this change, both on ascent to high altitudes and back again for descent.

POWER ASSIST CONTROLS

In the early days of flight, the aerodynamic controls (ailerons, elevators, rudder) were directly connected to the cockpit via cables, and the pilot had to use physical force to operate these controls. As the speeds of airplanes increased, the aerodynamic forces became larger and required more physical force from the pilot to operate the controls. With the advent of high-speed jet flight, these forces became too large for the pilot to overcome, and hydraulically boosted controls were introduced (much like power steering in your automobile). For the X-15, the power assist controls that gave force amplification to the pilot were effective; they were used by the pilots when the aerodynamic forces were high at the lower altitudes.

The power assist controls were used throughout by some of the pilots who did not use the conventional center stick and who only used the force amplification controls. The MH-96 also blended this control with the rocket controls, which were used when the air density was so low that the aerodynamic controls were ineffective because of the high altitude and resulting low dynamic pressure. It made the transition from aero control to rocket automatic. For use in future hypersonic aircraft, and in the Space Shuttle that actually followed, it simplified the piloting when flying in these varied regions of aerodynamic force. The X-15 demonstrated that airplanes in these regions, even while rapidly traversing from one region to another with high accelerations and decelerations, could be flown safely by trained pilots.

ROCKET CONTROLS IN SPACE

The use of rocket controls in flight was demonstrated earlier on the Bell X-1B airplane. Therefore, it was natural that rocket controls would be used for the X-15 as the only effective controls in space, where the aerodynamic forces are inadequate or nonexistent. These low-thrust rocket engines, using a monopropellant (hydrogen peroxide), provided useful control in space and have been used by the Space Shuttle in outer space.

ACCOMPLISHMENTS

All the design goals of the X-15 were met during its flight-test program, and some were surpassed.

The design maximum altitude and Mach number were both reached. The hypersonic research data obtained provided a rich database that confirmed the viability of hypersonic wind tunnel data as well as the usefulness of the limited theoretical analyses available at that time. The airplane proved to be a successful hypersonic vehicle, and the X-15 pilots performed admirably over an almost ten-year period. The program ended when the funding ran out and research experiments no longer justified the associated costs of the flights.

The flight region explored and extended the known range to M=6.7 and an altitude of 354,200 feet. The X-15 pilots explored this hypersonic range and provided data for future manned flights and for manned space vehicles flying from space through the atmosphere to landing, such as the Space Shuttle.

The new large RMI rocket motor performed well, providing the acceleration needed and with an operating efficiency of about 97 percent in support of obtaining mission data. There were no blowups in flight, and although the partial thrust use and subsequent restart capability were not reliable, the engine was able to position the airplane in the flight regions to be studied.

The MH-96 adaptive control system proved adequate and useful for stability on all three axes of flight. Some form of adaptive controls (controls that adapt automatically to the changing flight environment that was encountered during the flight of the airplane) have been used by high-performance aircraft in the fifty-plus years since the X-15.

All three control systems worked. The pilots preferred the power assisted controls over pure manual controls for use in the atmosphere, and the reaction rocket controls performed well in space and where the aerodynamic forces were insufficient. They have since been incorporated into the design of the Space Shuttle. The transition in use of the control system from space to the atmosphere where aerodynamic controls took over was easily effected.

The high-temperature material, Inconel X, maintained its strength as predicted at the high temperatures obtained in flight, and it supported the flight loads. This design approach, which allowed for thermal expansion of the hot structure while the cold understructure remained unstressed, was ultimately successful after the engineering team made a few corrections following initial hot flights.

The aero-thermodynamic analytical predictions *were* considerably higher than the actual measurements; analytics can now reliably use empirical data obtained from these flights. The research team also learned that the predicted high stagnation temperatures occurred where air could enter small gaps in wing construction, which then burned internal wires and structural features.

A ball nose instrument was attached at the extreme nose of the airplane and utilized Inconel X to withstand the high temperatures of hypersonic flight. This instrument, which provided angle of attack and angle of yaw data to the pilot, was necessary for flying and controlling the airplane at the high-speed and high-temperature conditions.

Replacement of ailerons was accomplished by using the horizontal stabilizer differentially deflected (i.e., right stabilizer angle increased while the left stabilizer angle decreased, and vice versa), providing satisfactory roll control and simplifying the knowledge of airflow conditions at the tail.

As a research airplane, the X-15 was also a useful platform for doing experiments at hypersonic speeds. Most important, the repeated and successful utility of this airplane over highly accelerated and decelerated flight from space to landing demonstrated that piloted aircraft are suitable for manned controlled return from space and for missions in the hypersonic regime.

The stable platform used to mate the X-15 to the B-52 malfunctioned at the start of the first X-15 government flight on March 25, 1960. Nevertheless, the flight took place. It was also test pilot Joe Walker's first X-15 flight. *USAF, Air Force Flight Test Center History Office, Edwards Air Force Base*

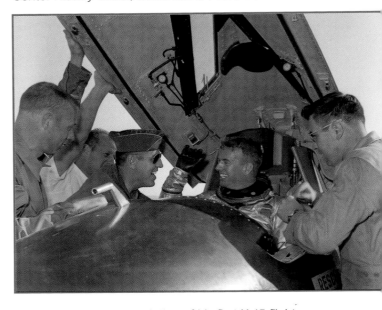

Joe Walker at the completion of his first X-15 flight, March 25, 1960. *USAF, Air Force Flight Test Center History Office, Edwards Air Force Base*

AFTERWORD

The 199th flight by Bill Dana in 1969 was the last for the X-15. The two of these revolutionary airplanes that still remained were readied for installation in national aviation museums after the completion of the X-15 research program.

As early as 1962, the Smithsonian Institution had requested an X-15 airplane for eventual display in Washington, D.C. The first X-15 was installed by the Smithsonian on May 13, 1969, in what was then known as Silver Hill and is now called the Garber Facility. It was moved to the Smithsonian's Arts and Industries Building in June 1969 and placed near the 1903 Wright Flyer. The Arts and Industries Building served as the National Air and Space Museum at that time. After being loaned out to the FAA and then to the NASA Flight Research Center for display, it returned to the Smithsonian to be installed in the new National Air and Space Museum in Washington, on the Mall, for its opening on July 1, 1976. It hangs there now in the

X-15-1 on display, October 15, 1958, after rollout at the North American factory, Inglewood, California. *USAF, Air Force Flight Test Center History Office, Edwards Air Force Base*

B-52 in flight with the X-15 mounted under the right wing, with the T-38 chase plane alongside. *USAF, Air Force Flight Test Center History Office, Edwards Air Force Base*

The X-15 in transit on a truck bed. Not everything was serious about the X-15 program; the mule is in case extra power is needed. *USAF, Air Force Flight Test Center History Office, Edwards Air Force Base*

X-15 test pilots Robert White and Joe Walker on parade. *USAF, Air Force Flight Test Center History Office, Edwards Air Force Base*

The X-15A-2 airplane went to the National Museum of the USAF at Wright-Patterson Air Force Base in Dayton, Ohio. A set of external tanks and a dummy supersonic combustion ramjet (scramjet) engine are part of that display.

The X-15-3 that crashed with Mike Adams was buried at an unknown location at Edwards Air Force Base.

The two B-52 carrier airplanes used by the X-15 program were reassigned by the Air Force after performing in the subsequent lifting body program at the NASA Flight Research Center.

The X-15 pilots continued with their careers: Neil Armstrong became famous as one of the first three men to land on the moon. Selected to be in the second astronaut class, he left the X-15 program, commanded Gemini 8, and on July 20, 1969, as commander of Apollo 11, became the first human to walk on the moon. His next position in NASA was deputy associate administrator for aeronautics at NASA headquarters. He left NASA to become professor of aeronautics at the University of Cincinnati, after which he served on the boards of several corporations. Neil Armstrong passed away on August 25, 2012.

Bill Dana became chief pilot at the Flight Research Center, then had progressively higher positions in Flight Operations, in F-18 research, and finally as chief engineer at the Flight Research Center, a position he held until his retirement in 1998.

Joe Engle was selected to become an astronaut in 1966 and performed as support crew on Apollo 10, then as backup lunar module pilot on Apollo 14. He commanded the Space Shuttle Columbia

Neil Armstrong in the cockpit of the X-15-3 prior to its first flight, December 20, 1961. *USAF, Air Force Flight Test Center History Office, Edwards Air Force Base*

and manually flew the reentry from Mach 25 through reentry and landing (the only time it was manually flown for an entire flight). His last flight in space was as pilot of Discovery in August 1985.

Pete Knight went to Southeast Asia and flew 253 combat missions in the F-100. He was test director of the F-15 System Program Office and piloted the airplane. He returned to Edwards Air Force Base as vice commander of the Flight Test Center and as an active F-16 pilot. He retired from the Air Force in 1982 and entered politics, rising to California state senator. He died on May 8, 2004.

Jack McKay retired from NASA in October 1971 and died on April 27, 1975, largely from complications from his X-15 crash.

Pete Peterson left NASA in 1962 and returned to the U.S. Navy, rising in rank after combat in Vietnam to be commander of the Naval Air Systems Command. He retired from active duty as vice admiral in May 1980. He died on December 8, 1990.

Bob Rushworth returned to the USAF after flying the X-15, and in the Vietnam conflict he flew 189 combat missions. He rose through the command ranks to become a general, and he retired as a major general from the position of vice commander of the Aeronautical Systems Division at Wright-Patterson Air Force Base. He died of a heart attack on March 17, 1993.

Milt Thompson remained with NASA after piloting the X-15, becoming chief of research projects. He then became chief engineer, a position he retained until his death on August 6, 1993. He wrote a wonderful book about his experiences flying the X-15, *At the Edge of Space*.

Joe Walker was helping obtain publicity shots of the XB70A while flying an F-104. Getting too close to the B-70 and caught in air currents between the two aircraft, he was killed in a midair collision on June 8, 1966.

Bob White continued in the United States Air Force. He became brigadier general and

X-15 landing on the lakebed. Note the high angle of attack on landing, a flight characteristic for hypersonic aircraft. *USAF, Air Force Flight Test Center History Office, Edwards Air Force Base*

Test pilot Robert White standing beside the X-15 on the occasion of rollout of the airplane. *USAF, Air Force Flight Test Center History Office, Edwards Air Force Base*

A moment of staged levity for some X-15 pilots. Note the camera in the foreground. *USAF, Air Force Flight Test Center History Office, Edwards Air Force Base*

commander of the Air Force Flight Test Center. He later became a major general and then chief of staff of the 4th Allied Tactical Air Force. He retired from the USAF in February 1981 and died on March 17, 2010.

Scott Crossfield, who left the NACA Flight Research Center to join North American Aviation to be a part of their X-15 design and flight-test team, ended his association with the X-15 program when the Air Force took it over. He then continued at NAA in many high-level and technical executive positions. He followed his NAA career with executive positions at Eastern Airlines and Hawker Siddeley Aviation. He then became a consultant to the House of Representatives Committee on Science and Technology. He lectured on aviation to many groups until his demise. It was after such a lecture at Maxwell AFB that he was killed in his Cessna 210 aircraft in a storm over Georgia while flying home on April 19, 2006.

At this writing, Joe Engle and Bill Dana are the only surviving X-15 pilots.

Collectively, the pilots who flew the X-15 airplane continued in their careers, flying for NASA in a research mode or for the military, where they progressed into positions of military leadership. Building upon their technical backgrounds and research piloting, they applied their work discipline to perform important responsibilities on behalf of the United States. They were talented men, driven and successful in their endeavors.

The X-15 remains the fastest and highest-flying manned airplane in history. The fact that no

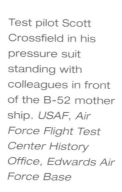

Test pilot Scott Crossfield in his pressure suit standing with colleagues in front of the B-52 mother ship. *USAF, Air Force Flight Test Center History Office, Edwards Air Force Base*

Joe Walker ready to enter the cockpit for his first flight on the X-15, March 25, 1960. This was the first government flight in the X-15 program. *USAF, Air Force Flight Test Center History Office, Edwards Air Force Base*

other manned hypersonic airplane has followed in its wake is a testimonial to the difficulty and severity posed by the hypersonic flight regime. The authors remain convinced that the future will see manned hypersonic flight for sustained periods in the atmosphere, a development that will rely on the data produced during the X-15 program on hypersonic aerodynamics, flight dynamics, structures, flight control, and pilot behavior. These hypersonic airplanes will be powered by air-breathing jet engines, not rocket engines. Such air-breathing engines will be supersonic combustion ramjet engines (scramjets), which have been under development since the 1970s and which are still a subject of intense research.

Indeed, on May 1, 2013, the experimental X-51, an unmanned hypersonic vehicle, achieved the longest duration sustained flight powered by a scramjet of over 300 seconds at speeds above Mach 5. The future of practical, environmentally safe, and economically feasible hypersonic manned flight still lies before us, and when that happens, the X-15 will indeed be the "Wright Flyer" of its kind.

▲ X-15 in flight. *USAF, Air Force Flight Test Center History Office, Edwards Air Force Base*

▼ The X-51 hypersonic research vehicle, powered by a supersonic combustion ramjet engine (scramjet). The X-51 is unmanned and is a waverider configuration for high lift-to-drag ratio. Its first flight was on May 26, 2010. Its fourth and final flight was on May 1, 2013, when it flew at Mach 5.1 for 240 seconds under scramjet propulsion, the longest air-breathing hypersonic flight to that time. *USAF*

INDEX

North American Aviation (NAA), 18–19, 20, 87, 139
Northrop, 77, 107, 109
Nortronics, 56

oblique shock waves, 52
Octave Chanute Award, 88

P-38, 90
P-40, 41
P-47, 30, 31
P-51 Mustang, 20, 30
P-59, 30, 31, 69
P-80, 11
Passman, Richard, 7
Peterson, Forest Silas "Pete," 63, 85, 93–94, 112–113, 138
Piasecki, 70, 80, 81
Pinecastle Field, 34
power assist controls, 133–134
pressure suit, David Clark, 123, 124

Reaction Motors, Inc. (RMI), 19, 32, 41, 61, 62
Republic, 87
Rice, Ray, 20
RMI rocket engine, 16, 19, 23, 134
rocket controls, 134
Rogers Dry Lake, 66, 68, 69, 72, 73, 101, 116, 128
Rome Air Development Center, 92
Rosamond, 73
rotational motion, 112
Rushworth, Robert A., 63, 71, 85, 96–99, 101, 112–113, 138

shock waves, 52
Shuttle Challenger, 104
Shuttle Columbia (STS-2), 104, 138
Shuttle Discovery (STS-27), 104, 138
Silver, 73
Skunk Works, 8, 11
Smith, Bill, 39
Smith Ranch, 73, 95
Society of Experimental Test Pilots, 95
sonic boom, 4–5, 8
sound, speed of, 5, 6
sound barrier, 4–5
Space Shuttle, 6, 14, 16, 20, 24, 82, 101, 104, 108, 109, 112, 123, 134, 138
specific impulse, 59, 61
speed brakes, 54, 104
speed records, 36, 42, 72, 96, 107
Spirit of St. Louis, 7
SR-71 Blackbird, 11, 123
Stabilization Augmentation System (SAS), 16, 56
Stack, John, 26, 28, 34
Stanley, Bob, 69
Storms, Harrison, 20, 62
Strategic Air Command, 73
supersonic combustion ramjet engine (scramjet), 107, 130, 137, 140, 141
supersonic flight regime, 29
swept wings, 38–39

T-38, 108, 137
T-38A, 77

Thompson, Milton O., 7, 63, 71–72, 85, 90, 96, 99, 101–102, 104–106, 116, 138
transonic regime, 16, 26

U-2, 11, 123
USS Caperton, 94
USS Enterprise, 94
USS Philippine Sea, 101

VF-20A squadron, 94
VF-154, 94
Volta Conference, 38

WAC Corporal rocket, 13
Walker, Joseph A. "Joe," 22, 38, 63, 81, 89–91, 105, 112–113, 116, 124, 130–131, 135, 137, 138, 140
water alcohol, 58, 61
wedge shape, 52–53
Westinghouse, 56
White, Al, 91
White, Robert M., 63, 82–83, 91–93, 112–113, 118–119, 137, 138–139
white ablator coating, 46–47
White Sands Proving Ground, 13
Williams, Walter C., 13, 22, 69, 84
Wolko, Frank, 41
Woods, Robert, 13, 30–31
Wright Aeronautical, 41
Wright Field, 29–30, 96
Wright Flyer, 14, 16, 136
Wright-Patterson AFB, 6, 92, 99, 107, 108, 137

X-1, 7, 8, 9, 13, 16, 19, 23, 28, 29–34, 36, 39, 42, 46, 48, 59, 61, 69, 73, 84, 86, 87, 123
X-1A, 13, 16, 23, 28, 35, 36, 38, 42, 46, 48, 61, 84, 87, 91, 123
X-1B, 35, 95, 134
X-1E, 91, 95
X-2, 7, 13, 16, 23, 28, 38–42, 46, 48, 69, 73, 84, 87, 123
X-3, 91
X-4, 86, 91
X-5, 33, 86, 91
X-15 (Jenkins), 7
X-15-1, 49, 94, 96, 98, 137
X-15-2, 12, 96
X-15-3, 48, 57, 66–67, 96, 111, 131, 137, 138
X-15A-2, 25, 28, 44–45, 50, 56, 59, 74–75, 96, 98, 107, 129, 132, 133, 137
X-20 Dyna-Soar, 105, 108
X-24B, 109
X-30, 87
X-51, 141
XB-70, 91
XB-70A, 138
XF-92, 86
XLR-11, 19, 32, 58, 59, 61, 62–63
XLR-99, 19, 21, 24, 58, 59, 61, 62, 63, 94
X-Planes, The (Miller), 7

Yeager, Charles "Chuck," 8, 9, 34, 36, 69, 87, 102

zero lift, 125
Ziegler, Jean, 41